Astronomers' Universe

For further volumes:
http://www.springer.com/series/6960

David A.J. Seargent

Weird Worlds

Bizarre Bodies of the Solar System
and Beyond

 Springer

David A.J. Seargent
The Entrance
NSW, Australia

ISSN 1614-659X
ISBN 978-1-4614-7063-2   ISBN 978-1-4614-7064-9 (eBook)
DOI 10.1007/978-1-4614-7064-9
Springer New York Heidelberg Dordrecht London

Library of Congress Control Number: 2013934105

© Springer Science+Business Media New York 2013
This work is subject to copyright. All rights are reserved by the Publisher, whether the whole or part of the material is concerned, specifically the rights of translation, reprinting, reuse of illustrations, recitation, broadcasting, reproduction on microfilms or in any other physical way, and transmission or information storage and retrieval, electronic adaptation, computer software, or by similar or dissimilar methodology now known or hereafter developed. Exempted from this legal reservation are brief excerpts in connection with reviews or scholarly analysis or material supplied specifically for the purpose of being entered and executed on a computer system, for exclusive use by the purchaser of the work. Duplication of this publication or parts thereof is permitted only under the provisions of the Copyright Law of the Publisher's location, in its current version, and permission for use must always be obtained from Springer. Permissions for use may be obtained through RightsLink at the Copyright Clearance Center. Violations are liable to prosecution under the respective Copyright Law.
The use of general descriptive names, registered names, trademarks, service marks, etc. in this publication does not imply, even in the absence of a specific statement, that such names are exempt from the relevant protective laws and regulations and therefore free for general use.
While the advice and information in this book are believed to be true and accurate at the date of publication, neither the authors nor the editors nor the publisher can accept any legal responsibility for any errors or omissions that may be made. The publisher makes no warranty, express or implied, with respect to the material contained herein.

Printed on acid-free paper

Springer is part of Springer Science+Business Media (www.springer.com)

For Meg and Elliott

Preface

In my previous "weird" volumes—*Weird Astronomy* and *Weird Weather*—the principal topics of interest were largely observations that appeared anomalous or that in some way did not "fit" with accepted ideas. The present volume has a slightly different emphasis. From the point of view of Earth, the worlds beyond our own all have aspects that seem "anomalous" or "weird" to us, not because they fail to accord with accepted ideas but simply because they harbor phenomena which lie outside the domain of our normal experience. If some of the things that are regular aspects of the scenery on Mars or Mercury—not even to mention Titan or, still less, planets of other suns—were to occur on Earth, they would make the "weird" events and observations discussed in my previous volumes pale into insignificance! What if the Sun were to halt in the sky, reverse on its track for a while, and then continue as "normal" or rise in the west and set in the east? What if jets of carbon dioxide suddenly erupted from beneath the polar ice, shooting fountains of dust high into the air; or liquid methane rained from the skies; or volcanoes shot fountains of lava so high as to effectively reach the edge of space? What should we think about ice remaining solid at a temperature equivalent to that of white-hot metal or diamond dust falling through a liquid realm that is neither atmosphere nor ocean? Could we even imagine the possibility of finding a form of living organism that would explode in a ball of flame if exposed to the air? Who could conceive that a stone dropped from the top of a cliff could take a quarter hour to reach the bottom? And what more bizarre sight could there be than a moon progressing through all its phases as it crossed the sky between moonrise and moonset—especially if it also went "backward" from west to east?

Yet all of these things are commonplace on other worlds of which we already have some knowledge. Add to this the prospects of simple life in underground oceans on moons and even asteroids, comets that practically strike the Sun as they whip around it at over a million miles per hour, planet-like bodies wandering through the cosmic night of interstellar and maybe even intergalactic space, planet-sized diamonds and similarly proportioned balls of steel and we can agree that some pretty weird things lurk in this Universe. Through the pages of this book, we will look at some of them.

Following the pattern of the previous two volumes, the reader will find several "Projects" within these pages. Most of these are simple observing exercises. Some can be included in astronomy club open nights or suggested as exercises that junior club members might like to try. Either way, they emphasize the fact that astronomy is first of all an observational science and one in which the amateur having only modest equipment can participate and, by participating, come to share something of the excitement of our increasing knowledge of the weird denizens of this wonderful Universe.

The Entrance, NSW, Australia David A.J. Seargent

About the Author

David A. J. Seargent holds an M.A. and a Ph.D., both in Philosophy, from the University of Newcastle, New South Wales, where he formerly worked as a tutor in Philosophy for the Department of Community Programs/Workers' Educational Association external education program. He is also an avid astronomer and is known for his observations of comets, one of which he discovered in 1978. Together with his wife Meg, David lives at The Entrance, north of Sydney, on the Central Coast of New South Wales, Australia. He is the author of four published astronomy books: *Comets—Vagabonds of Space* (Doubleday, 1982), *The Greatest Comets in History—Broom Stars and Celestial Scimitars* (Springer, 2008), *Weird Astronomy—Tales of the Unusual, Bizarre, and Other Hard to Explain Observations* (Springer, 2010), and *Weird Weather: Tales of Astronomical and Atmospheric Anomalies* (Springer, 2012). Currently, he is the author of a regular column in *Australian Sky & Telescope* magazine.

Acknowledgments

Many people have, in a variety of ways, helped to encourage my interest in the subjects that form the theme of this book. To all these folk, some of whom are no longer with us, I extend my thanks.

Coming to more recent times, I would like to thank my wife, Meg, for her support and the staff at Springer Publishing, especially Ms. Megan Ernst, Ms Maury Solomon and Mr. John Watson for their encouragement.

Last, but by no means least, I would like to thank all those who have developed the Internet into the store of accessible information that we have today. It certainly makes the ferreting out of obscure pieces of information an easier task than it was not so very long ago!

Contents

1 Four Rocks Near the Sun 1

2 Giants of Gas and Ice 65

3 Asteroids, Dwarf Planets and Other Minor Bodies 121

4 Moons Galore! 175

5 Titan: The Weirdest World in the Solar System?! 241

6 Weird Worlds Far Away 265

Name Index 301

Subject Index 305

1. Four Rocks Near the Sun

That tiny region of space lying between the Sun and the inner Main Belt of asteroids is the most familiar to us. No surprise in that of course; after all, this is where our own blue Earth resides. Our home is, to use a phrase made popular by a science fiction comedy of late last century, the "Third Rock from the Sun". So before venturing any further into the Universe, it will be well to take a look around this cosmic backyard of ours and the other three "Rocks" that share it with us.

Is there anything especially "weird" or odd about these?

Let us take a look and find out.

Mercury

On the face of it, there does not appear to be anything particularly weird about barren little Mercury. If this is how you feel, then prepare to have your opinion challenged as you read the following pages!

Let us begin by looking at a few features of this world which might be rated as distinguishing, even if not all of them are very complimentary.

To begin, it is the smallest Solar-System planet that is still classified as what we might call a "planet without prefix", i.e. not a "minor planet", "dwarf planet" of some similar diminutive term (actually, it *has* recently been given a prefix as we shall see a little later, but that involves one of its few weird features, not its small size). Mercury has a diameter of just 3,050 miles (4,880 km). Putting this in familiar perspective, the distance between New York and San Francisco is around 2,600 miles and that between Sydney and Perth, a little over 2,000. So Mercury would just cover the USA and Australia!

2 Weird Worlds

FIG. 1.1 Mercury, showing Kuiper crater (Credit: NASA)

It is also distinguished by being the planet closest to the Sun with a mean distance of only 36 million miles (about 57.6 million kilometers), compared to just over 93 million miles (about 150 million kilometers) for Earth. In terms of the unit used to measure distances within the Solar System where the mile and kilometer are too small and the light year too large, Mercury has a mean solar distance of 0.39 Astronomical Units (AU). One AU is the Earth's mean distance from the Sun, so the mean radius of Mercury's orbit is therefore thirty nine hundredths that of Earth's. Mercury is no place for thermophobes!

In consequence of its small distance from the Sun, Mercury is also the fastest of the Sun's planets and the one with the shortest year. It whips around its orbit at an average speed of 29.92 miles (47.87 km) per second and completes one full revolution of its orbit—one Mercurian year—in a mere 87.97 terrestrial days. It may not be too polite to mention this, but a 30-year-old earthling has already reached 124 in terms of Mercurian years! (Fig. 1.1).

Of all the planets of the Solar System, Mercury moves along the orbit furthest removed from a true circle. In more technical terms, its orbit has the highest eccentricity of all the major planets; 0.206, compared with 0.006 for Venus and 0.016 for Earth. Even Mars, renowned for its eccentric orbit, only scores 0.093. It is true that Pluto clocks in with a greater eccentricity at almost 0.26, but as this object has now joined the ranks of the planets-with-prefixes (viz. a dwarf planet) this is no longer a challenge to Mercury. The rights and wrongs of Pluto's "demotion" (as some folk see it) fortunately do not concern us here!

Keeping up the "extreme" standard, the angle at which Mercury's orbit is inclined to the ecliptic (essentially the plane of Earth's orbit) is more than twice as large as the next largest (that of Venus)—once again excluding dwarf planet Pluto. The inclination of Mercury's orbit is just over 7°, that of Venus a little less than 3.5°, Saturn just under 3 and the rest less than 2. By contrast, the planet's axial tilt is just 0.03°—about 100 times smaller that of the next lowest in the Solar System; that of the giant planet Jupiter!

Partially because of the eccentric nature of its orbit and partially because of its lack of substantial atmosphere (see below), temperatures, although always high when the Sun is above the horizon, do nevertheless vary quite markedly. The average temperature of Mercury is 169.5 °C, but the range extends from a frigid −173 °C at the floors of polar craters (more about these locations later) to 427 °C at the subsolar point when the planet is at perihelion i.e. when it reaches its closest point to the Sun. This temperature is easily high enough to melt lead, but when Mercury is at aphelion (i.e. the furthest point in its orbit from the Sun), subsolar temperatures are fully 150 °C lower at "only" 277°. Lead would stay solid!

Mercury is also uncommonly dense. In fact, at 5.43 times that of water its density is only slightly less than that of the densest member of the Sun's family of planets; our very own home world Earth, which has a density 5.52 times that of water. Indeed, when the effect of gravitational compression on both planets is factored in, Mercury actually emerges as the potentially denser of the two, having an "uncompressed" density of 5.3 as against 4.4 for Earth. We will say more about its high density later.

Because this planet has such an eccentric orbit, plus its location inside the orbit of Earth (in the sense that its orbit lies between that of Earth and the Sun), its distance from Earth can vary considerably and this is partially responsible for another of Mercury's extremes, namely, the wide variation in its brightness. The planet's apparent brightness—that is to say, how bright it appears from our perspective on Earth—goes through a greater range than that of any other planet, from approximately that of distant Uranus at faintest to the brilliance of Jupiter at its brightest. This range is not immediately obvious however. Being inside the orbit of Earth, Mercury goes through phases similar to those of the Moon, a phenomenon first observed by Giovannit Zupi as long ago as 1639. At its faintest, the planet is (somewhat ironically) about as close to Earth as it can be, but is located in our sky very close to the Sun and appears as nothing more than a very thin crescent. At least, that's how it *would* appear if we could even see it in the Sun's glare! At brightest, it is on the far side of the Sun and presents a fully illuminated, albeit very small, disk. However, once again, it is very close to the Sun in the sky and not at all easy to see in spite of its greater brightness. The planet is easiest to see when it is neither far beyond the Sun nor close in front of it, but more or less side on to it from Earth's perspective. At these maximum elongations, it is observable in deepening twilight as a bright star-like body with a pinkish coloration. This is how the planet typically appears to us, because these are the occasions when it is most readily visible. Even so, it is never located more than 28.3° from the Sun and, moreover, these maximum elongations only occur south of the celestial equator. Mercury favors the Southern Hemisphere, where in can sometimes be seen in a dark sky. For Northern observers, it is never quite clear of twilight and for both hemispheres it sinks closer to horizon haze as the skies darken. There is a story that Copernicus never saw it because of rising mists near his home. Like many such "legends", this one is probably not correct.

The planet's passages more or less behind or in front of the Sun are known as *conjunctions*. When Mercury is behind the Sun—on the far side of its orbit from Earth—the conjunction is termed a *superior* conjunction and when the planet passes between Sun and Earth at the near side of its orbit, we have an *inferior* conjunction. Of course, Mercury seldom passes either directly behind or in front of the Sun, but merely reaches a minimum angular elongation from it. Passage directly behind results in an occultation of the planet by the Sun. These are, for obvious reasons, unobservable by ordinary means. On the other hand, on the rather rare occasions when Mercury *transits*, or passes directly in front of the Sun, it is observable as a small black speck drifting slowly across the brilliant face of our star. The first such transit to be observed was by Pierre Gassendi in 1631 following a prediction made by Kepler. Many have been observed in the interim; the next one being due on May 19, 2016 and the following on November 11, 2019. Incidentally, the terms "superior" and "inferior" carry no judgmental sense as to the quality of the different types of conjunction. These terms are also applied to the planets themselves, once again without any judgmental overtone. Mercury and Venus are referred to as the "inferior" planets because their orbits are inside (Sunward of) Earth's. Planets beyond Earth are "superior" because of their location beyond Earth, not because they are in some sense 'better' than the other two.

Although not the easiest planet to observe in detail, Mercury's brightness at greatest elongation makes it briefly conspicuous to naked eyes (the Copernicus tale notwithstanding!) and the planet was one of the "wandering stars" known to the Ancients, with references to it found in records dating back to at least 1,000 years before Christ. Nevertheless, some of the earliest stargazers do not seem to have realized that the bright "star" sometimes seen in the breaking dawn and the one seen on other occasions in the evening twilight were one and the same. At least, that appears to be the conclusion drawn from the fact that Greek astronomers living before the fourth century BC gave it two names; Apollo in the morning skies and Hermes in the evening. As "Hermes" is never seen on the evenings when "Apollo" graces the dawn—and vice versa—it is strange that they took so long to jump to the correct conclusion. One might wonder if they really did believe the two were separate objects. Or, if that was the "official" position, how many early Greek astronomers held their own private ideas on the subject! (There is a story that Pythagoras suspected the identity of these supposedly two objects back in the fifth century BC. Whether it was actually Pythagoras or one of his students might be debatable, as one of the rules of the Pythagorean "school"—"cult" would probably be a more accurate term!—was to ascribe to the Master any discovery made by any of its members. On one occasion, a Pythagorean student dared take credit for his own discovery … and was later found drowned at sea. A strong deterrent for such precocity, one might imagine!)

In any event, "Hermes" won the day eventually, albeit by his Roman name, *Mercury*. Today, the names *Apollo* and *Hermes* have been redirected to two Earth-approaching asteroids.

Despite its relatively easy naked-eye visibility—given a clear horizon and freedom from rising mists—Mercury is not a good object for telescopes and early telescopic astronomers could make out little detail on its surface. At one time, it was thought that (like its brilliant neighbor Venus) the planet was perpetually shrouded in cloud; however the problem lies more with our own atmosphere than with any gaseous mantle surrounding Mercury. The big problem is that Mercury is always low when the sky is anywhere near dark. And low means that we are seeing it through the greatest depth of our atmosphere, with all the legion of problems that this brings. Add to this the planet's small diameter and the fact that it is not observed when at full phase and it is no surprise that telescopic scrutiny of Mercury brought less than startling progress. The clearest telescopic observations tended to be those made in full daylight when the planet was high in the sky. Incidentally, the familiar pinkish color pretty much disappeared in these daytime views, indicating that this owes more to the low altitude of the planet during typical sightings than to the true color of its surface.

Nevertheless, despite all the problems, subtle details were seen on Mercury as early as the year 1800 and some 80 years later, the first accurate map was drawn by G. Schiaparelli. Schiaparelli noted that the face of the planet looked the same every time he observed it, rather as the same features on the Moon are seen over and over again from our terrestrial perspective. Clearly, just as the Moon has the same face turned toward us, so the same must apply to Mercury, presumably (so Schiaparelli reasoned) for the same reason, i.e. Mercury's rotation is tidally locked in a way similar to that of the Moon. Of course, unlike the Moon, it is not locked to the Earth but to the Sun, yet the observational result is the same. We only see one face of the planet. More will be said about this anon.

During the 1930s, planetary astronomer E. M. Antoniadi constructed another map of the planet from his observations made with the 33-in. (84-cm) refracting telescope at Meudon Observatory. This became the "canonical" one for later observers.

Despite giving them romantic sounding names such as *Solitudo Hermae Trismegisit* (Wilderness of Hermes the Thrice Greatest), the features that Antoniadi recorded appeared as nothing more than vague patches. But that is in no way to belittle his effort. Far from it. Recording *anything* on Mercury requires skill and patience beyond the normal call of duty!

Project 1: Mercurian Markings

The well-known and skilled British amateur astronomer Patrick Moore is on record as saying that he "glimpsed" the main markings (he referred to them as "patches") on Mercury with the aid of a 6-in. (15-cm.) refractor. Although he does not say where he was observing, presumably it was from somewhere in the British Isles where Mercury is less well placed than from the Southern Hemisphere (unless he was observing in broad daylight.)

All this raises the question "What size telescope is required to see the main markings recorded by Antoniadi?" Those with the best chance of answering this question are observers living south of the Equator who have had a good deal of practice observing faint features on other planets, but the challenge is open to anyone with a telescope. Can *your* telescope see the markings? Try observing in bright twilight when Mercury is at maximum elevation. The image will be less affected by atmospheric turbulence and there may also be some advantage in so far as the planet will not appear as brilliant in the eyepiece (although that is not as great a problem for Mercury as it is for Venus). You may like to try observing the planet in full daylight, when it is high in the sky, however this should be done with the greatest of care and ONLY when the Sun is hidden behind a building (and sinking further behind it all the time) or if your telescope mount is equipped with some form of positioning device; whether computerized "go to" or old fashioned setting circles. **NEVER sweep for Mercury in the daytime**. It is always too close to the Sun for safe sweeping and the risk of having the Sun enter the eyepiece field and burning out an eye is just too great!

8   Weird Worlds

Not surprisingly the markings on the planet appeared to be permanent features, although their true nature could only be guessed at before the advent of space probes. By the time of Antoniadi's observations however, any notion of atmospheric clouds had gone and Mercury was thought to be lacking any kind of atmospheric envelope whatsoever. Nevertheless, in 1950, astronomers announced the apparent discovery of an extremely thin atmosphere and more recent research has confirmed the presence of what has been termed a "tenuous surface-bounded exosphere" consisting of 42 % molecular oxygen, 29 % sodium, 22 % hydrogen, 6 % helium and 0.5 % potassium together with a mixture of trace amounts of nitrogen, water vapor, magnesium, argon, xenon, neon and krypton together making up the remaining 0.5 %.

The planet is too small and too hot to hold a "true" atmosphere—even a very tenuous one. Because its constituent atoms are constantly being lost—swept away by the relentless radiation of the nearby Sun—the atmosphere requires constant sources of replacement. The very light hydrogen and helium atoms most probably come from the same source that eventually sweeps them away again; the Sun. A somewhat surprising discovery of space probes is that Mercury has a magnetic field. Although measured at just 1.1 % the strength of Earth's (which is also pretty weak as magnetic fields go—a dressmakers' magnetic does far better!) it is enough to diffuse the incoming solar wind of hydrogen and helium atoms, temporarily holding them until they later escape back into space. Some helium may also be released by the radioactive decay of elements in the crust of the planet itself. Sodium is probably sputtered off Mercury's surface by the intense solar radiation. Although it has a magnetic field, this is too "leaky" to effectively shield the planetary surface from energetic particles from the Sun. During the second flyby by the *Messenger* spacecraft on October 6, 2008, twisted bundles of magnetic field as wide as one third Mercury's radius were discovered connecting the planet's magnetic field with that of the stream of charged particles forming the Solar wind. These "magnetic tornadoes" form when the Solar wind, carrying its own magnetic field, blows past the planet and the fields of both planet and Solar wind twist up into vortex-like structures which can effectively act as "tubes" through which the Solar wind blows right down onto the planet's surface and sputters most of the sodium atoms that are observed in the tenuous atmosphere. Incidentally, these magnetic swirls have their familiar analogues in the peculiar wave-like clouds visible from time to time in Earth's atmosphere, except that here the eddies are created by the interaction of two bodies of air, not a clash of planetary and interplanetary magnetic fields. Similar waves (officially known as Kelvin-Helmholtz waves) occur wherever there is a boundary between two moving fluids.

Some atmospheric sodium, along with potassium and calcium, is also thought to come from the vaporization of surface rock as it is struck by micrometeorites. Water vapor might be formed from the combination of solar-wind hydrogen atoms and oxygen sputtered from surface rock but it may also come from the very slow sublimation of deposits of ice. Yes, that is correct. Mercury appears to have significant quantities of ice!

Now that is really weird. Ice on a planet where the sunlight can be hot enough to melt lead! Yet, radar observations in the early 1990s found highly reflective patches near the planet's poles and water ice appears to be the most likely explanation.

Of course, there is no real analogy with Earth's polar ice caps. If the interpretation of these observations is correct, the ice of Mercury resides in perpetual shadow on crater floors. The reason why shadows are long enough and sufficiently persistent on the floors of polar craters is down to a remarkable fact about Mercury's axial tilt; just 0.03°, the smallest of all the planets, as noted earlier. That means that the Sun is never much more than 2 minutes of arc, or thereabouts, above the horizon at the poles! Because of this unusual feature, significant amounts of ice can remain stable on this traditionally roasting hot first rock from the Sun!

The ice itself is thought to have been delivered by impacting comets and it is possible that some of the planet's atmospheric water vapor comes from this source as well. Scientists were quite surprised by the unexpected quantity of dissociated products of water vapor (ions such as O^+, OH^- and H_2O^+) in the region of space surrounding Mercury and surmised that these have either been blasted from the surface or swept from the exosphere by the solar wind. The transitory, but ever replenishing, nature of the Mercurian exosphere makes it appear (in these respects at least) more akin to the coma of a comet than to the relatively stable envelopes of Venus, Earth and Mars. We will see in a little while how apposite this comparison really is!

As mentioned above, since the time of Schiaparelli's mapping in the 1880s, astronomers believed that Mercury's rotation about its axis was gravitationally locked to the Sun, so that its "day" equaled its "year" and one hemisphere remained forever turned toward the Sun, roasting in eternal sunlight, while the dark side froze near absolute zero in everlasting night. As Patrick Moore stated in the early 1960s "It is not correct to term Mercury 'the hottest planet'; more properly it is 'the hottest and coldest planet'." As we will later see, Venus has stolen the "hottest (Solar System) planet" title from Mercury thanks to its extreme greenhouse effect. Nevertheless, Moore was essentially correct about the innermost planet's extreme temperature range. But the range is between the Sun-exposed open surface and the perpetual darkness of the floors of some polar craters. He was quite wrong about this radical temperature division existing between hemispheres. That is because, contrary to the accepted wisdom at the time he wrote these words, it turns out that Mercury is not tidally locked with the Sun after all. This means that both hemispheres of Mercury get some relief from the relentless heat and cold. The night side never gets as cold as Moore and his contemporaries thought and the Sun really does set on Mercury's daylight hemisphere. Nevertheless, this does not make the planet more homely. The reality of Mercury's rotation is even weirder than initially believed!

Observations of the planet using radio telescopes around the middle years of last century gave the first indications that its nighttime side was not as frigid as expected. If Mercury had a decent atmosphere, that would be no mystery, but with just the barest trace of a gaseous mantle, transfer of heat from the daytime to the nighttime sides would require impossibly high winds. The mystery was solved when Doppler Radar observations using the giant radio telescope at Puerto Rico in 1965 suggested a rotation period about two-thirds that of the orbital, i.e. 59 days in round figures. Astronomer G. Colombo proposed that the planet was tidally locked into a 3:2 spin-orbit resonance, i.e. rotating about its axis three times for every two trips around the Sun. This was subsequently confirmed by data from the *Mariner 10* spacecraft in 1974–1975. Because the planet's orbit is—as we earlier saw—very eccentric by planetary standards, this resonance remains stable.

The resonance has a peculiar effect on the length of the Murcurian day. We might suppose that, because the planet turns on its axis once every 59 days, the length of its day—from one sunrise to the next—is equivalent to that period of time. But not so! Because the planet rotates on its axis three times for every two orbits of the Sun, the actual period from one sunrise until the next is as long as 176 days, approximately two whole Mercurian years. At the time the planet experiences its strongest solar tide—i.e. when it is at perihelion—the Sun almost stands still in Mercury's sky. In fact, as viewed by hypothetical (presumably asbestos-coated!) dwellers at certain suitable places on the planet's surface, the Sun would even be seen to reverse its course across the sky for a brief period at the time of perihelion! This is because, at its minimum distance from the Sun, Mercury's orbital speed briefly out paces its velocity of rotation. A Mercurian sunrise would be truly weird at such times. At a suitable location on the planet's surface, an observer would see the Sun start to rise above the horizon, slow to a halt, and then drop back down again out of sight. Sunrise and sunset in quick succession at one point on the horizon. It would not remain "set" for long however. In a little while, the Sun reappears and resumes its "normal" course across the heavens, eventually setting, this time behind the opposite horizon, for a second time within a single Mercurian day!

An earlier generation of astronomers wrongly concluded that the planet's rotation was tidally locked because, when most favorably placed for observation from Earth, it is nearly always at the same point in its 3:2 resonance and the same surface features are therefore visible. This is the product of an odd coincidence. The orbital period of Mercury is close to half that of its so-called synodic period, i.e. the period between two successive conjunctions with the Sun as seen from Earth. This means that when the planet emerges from conjunction and reaches optimum visibility, it has made two whole orbits of the Sun and, given the spin-orbit resonance, has the same face turned toward us as on all previous similar occasions. The favorable circumstances during which Mercury was well enough placed for detailed observations to be made occurred during alternate orbits of the planet; the intermediate ones—when it had its other face turned toward the Sun and potentially visible from Earth—were less favorable and therefore less conducive to satisfactory observations.

The relatively high eccentricity of Mercury's orbit is something of a curiosity in its own right. One might think that an object orbiting so close to the Sun would have had its orbit "ironed out" into something just about as close to a perfect circle as is likely to be found in Nature. As we have already noted, its closest neighbor, Venus, has an orbital eccentricity of just 0.006, the smallest of all the Sun's planetary retinue and a far cry from the 0.2 of Mercury. It has been suggested that Mercury's orbit may be the result of an early collision with a protoplanet (for which there are other, albeit unproven, indications as we shall see), however computer simulations suggest that such a dramatic explanation for its odd orbit is not necessary. According to orbital simulations run by A. Correia, C. M. Alexandre and J. Laskar, the planet's orbit varies chaotically from having an eccentricity of nearly zero to more than 0.45 over periods of millions of years thanks to the gravitational perturbations of the other planets. These authors argue that this is also likely to explain Mercury's odd 3:2 spin-orbit resonance. A resonance of 1:1 is more usual and is what one would normally expect for a planet, however the more exotic 3:2 has a higher chance of arising if an orbit is unusually eccentric and, in Mercury's case, probably arose during a period of high eccentricity. Moreover, further simulations by Laskar, together with M. Gastineau, indicate that a resonant orbital interaction with Jupiter might so increase the eccentricity of Mercury's orbit that the planet will eventually cross the orbits of Venus and Earth and have a 1 % chance of actually colliding with one of these planets within the next five billion years. A chilling thought, but one which we need not lose any sleep worrying about for a long time to come!

Mercury has been visited by two spacecraft; *Mariner 10* which mapped about 45 % of its surface during 1974–1975 and *Messenger*, launched on August 3, 2004 and finally achieving orbit around the planet on March 17, 2011, after having made earlier flybys on January 14 and October 6, 2008. Not surprisingly, these revealed a planetary surface dotted with impact craters, basins and plains, not unlike that of the Moon. Clearly, this little world has suffered some awful impacts in the distant past. One impact feature known as the *Caloris Basin*, one of the largest known impact craters in the entire Solar System, has a diameter of some 969 miles (about 1,550 km). The impact that gouged out *Caloris* was so powerful

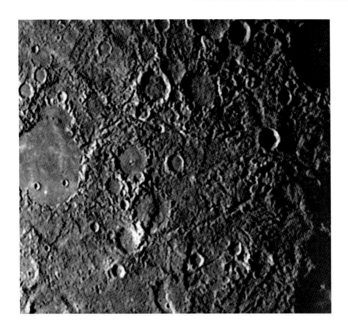

FIG. 1.2 Weird terrain on Mercury (Credit: NASA)

that it triggered eruptions of lava creating a ring about one and a quarter miles high concentric with the crater rim itself. Not only that, on the exactly opposite point of the planet lies a region so odd that it has simply been dubbed "Weird Terrain"; a jumble of odd hills and ridges that is thought to have arisen, either from ground shock waves travelling right around the planet and converging at the antipode of the impact site or else by the convergence of actual ejecta thrown up by the impact. Or maybe even, a combination of both. A feature not seen on the Moon is the presence of long narrow ridges running across the planet's surface. These are thought to have formed as Mercury's core and mantle cooled and contracted after the crust had already solidified. Speaking of *Caloris Basin*, recent *Messenger* observations show that some parts of the floor of this giant crater are higher than its rim; an unexpected discovery about which more will be said in a little while (Fig. 1.2).

Which raises the issue of the planet's core!

This was something that was not expected! From *Mariner 10* data and Earth based observations, the core of Mercury is estimated to account for 42 % of the planet's volume and takes up approximately 85 % of the planet's radius. Compare this with

Earth's 17 % core (itself considered pretty large) and it will be appreciated that this oversized core is the weirdest physical aspect of this small planet. Mercury may look rather like the Moon, but underneath that cratered crust, this little planet is very different. Not only is its core proportionally in a class of its own amongst the Sun's planetary family, but it also possesses the highest iron content of any of the major planets.

Overlaying this core is, planetary scientists believe, a solid layer of iron sulfide which in its turn is overlain by a thin shell of silicate mantle and crust.

The most widely held explanation for the outsized and unusually iron rich core of Mercury proposes that the planet was once about two and one quarter times its present size but, early in the lifetime of the Solar System, it was struck by a planetesimal about one sixth of its mass and several 100 miles in diameter. The impact is thought to have stripped away much of the original mantle and crust, leaving behind the iron core and not much else. After a time, some of the crust and mantle material fell back to form a rocky sphere around the core and the Mercury that we know today took form. A somewhat similar impact event is widely held to have given Earth its oversized moon, except that in this instance, the impact was a grazing one that left considerably more of proto-Earth behind than just its core.

Although this hypothesis is widely held, it is not without competitors. One alternative theorizes that Mercury formed quickly from the solar nebula before the Sun's energy output had stabilized and our star reached the Main Sequence. This scenario also postulates a larger initial Mercury—about twice its present size actually. As the protosun contracted and switched on, temperatures on Mercury rose to such extremes (between about 2,230 °C and 9,700 °C) that the planet's surface rock turned to vapor and simply blew away in the Solar wind, leaving behind a much depleted planet still possessing the core of a much larger world; a core which had become disproportionally large for the planet's greatly reduced diameter.

Yet another hypothesis suggests that the early nebula from which the Sun and its planets formed caused sufficient drag on solid particles that, at the relatively small distances at which Mercury formed, a large percentage of particles of lighter substances got swept out of the accreting material, leaving an overabundance of heavy ones (such as particles of, or rich in, iron) to coalesce together and eventually form a planet more than normally endowed with iron and, in consequence, possessed of a disproportionally large core of this metal.

The jury is still out on which hypothesis is most likely correct or, indeed, if yet another is required.

Although once assumed geologically inactive for most of its existence, recent findings from the *Messenger* spacecraft indicate a more interesting Mercurian history. Observations of craters reveal that many of these have tilted since their formation and it is thought this is most readily explained in terms of deformation caused by changes deep within the planet. One extreme case concerns the giant *Caloris Basin*. As noted earlier, parts of the floor of this feature are higher than the rim; something which certainly would not have been true during the period immediately following the impact. It seems clear that these elevated regions could only have been pushed upward by forces originating deep within the planet. Taken all together, this evidence of surface deformation strongly implies that Mercury was subject to tectonic forces for a long time, although explaining these forces is not proving to be easy. Maria Zuber of MIT suggests that convection may have led to mass circulating in the interior of the planet, but wonders how this could have proceeded given Mercury's extremely thin mantle. Something odd and interesting seems to have been happening here, although just what that could have been is not at all clear.

The Planet with a Comet-Like Tail!

So we can see from the above overview of this diminutive planet that, far from being an unexciting little rock near the Sun, it actually possesses some very interesting and—dare we say it?—even "weird" features. Yet thus far we have really only looked at what we might call its physical nature. We have left until the end an odd feature that *observationally* places Mercury in the "weird" category. This planet sports a tail like that of a comet! (Fig. 1.3).

Fig. 1.3 This is not a comet: It is Mercury with a tail (Credit: NASA/STEREO)

It has been known for some time that a tail of sodium atoms extends away from Mercury's atmosphere in the anti-solar direction, accelerated to escape velocity from the planet's gravity by interaction with the Solar wind. Yet, the full realization of this feature's extent awaited the launch of the twin *STEREO* Sun-monitoring spacecraft in 2006. The tail was spotted in on-line data from the *STEREO* project by Australian medical researcher and astronomy enthusiast Dr. Ian Musgrave and it was quickly realized that the intensity of the feature in these white-light images was significantly higher than could be accounted for by sodium emission alone. More than sodium is clearly being accelerated out into the tail. Moreover, the tail has also proven to be longer than at first thought. Ground based observations from the McDonald Observatory in Texas in 2008 (made in the light of glowing sodium) imaged the tail flowing out some 1.5 million miles from the planet (we might almost say "comet"). Tail lengths of 2° have been imaged from the ground. Clearly, there is more to be learned about this odd Mercurian feature.

Back in 2006, sodium source regions were identified at high Mercurian latitudes by the 3.7-m telescope on Mt. Haleakala and these are probably regions where the Solar wind impinges most strongly on the planet's surface. From observations made at Kitt

Peak, it was calculated that between 10 and 20 g of sodium are swept into the tail each second by the Solar wind, representing just 1–10 % of the total amount of sodium estimated to be released from the planetary surface.

Earlier, I said that Mercury is a planet without prefix (like "minor-" or "dwarf-" or whatever) but then modified this statement by saying that a prefix of a different kind has now been applied to the planet by some. Now is the time to reveal what this prefix is: It is "comet-". Mercury has been called a "comet-planet" because of the existence of this comet-like tail. The similarity extends beyond the mere existence of a tail however. As we have seen, Mercury has an atmosphere that has as at least as many features of a comet's coma as it has of the atmospheres of planets like Earth or Mars. It even has some slowly sublimating ice contributing photo-ionized products of water vapor to surrounding space—just like a comet! But, of course, it is not a comet. Simply a planet that has some of the features of a comet; yet another instance of nature refusing to draw the nice neat boundaries that we humans love to impose upon her!

So this is Mercury, the first of the Sun's four inner rocks. And what a strange little planet it is too! A world with an oversized core, where the Sun periodically stands still and then reverses in the sky and yo-yos at sunrise for suitably placed locations. A world with an atmosphere that in certain respects resembles the coma of a comet and which completes the similarity by sporting a long flowing tail. Once thought to be nothing more exciting than a ball of rock, the Solar System's smallest genuine planet turns out to be a fascinating world indeed!

Venus

The next rock outward from the Sun is the brilliant Venus, a spectacle brighter than any other regular denizen of the sky excepting the Sun and Moon. At its brightest, it gleams at an impressive magnitude –4.9 in the heavens and can be seen without too much trouble with the naked eye in full daylight (where, by the way, it has triggered not a few UFO scares!). Unlike Mercury, its faintest magnitude is still very impressive, still outshining all the other planets at magnitude −3.8. Also differing from Mercury, Venus is actually faintest at "full" phase when it is furthest from Earth and shows a disk just 9.7 seconds of arc in diameter. Nearest at crescent phase, its disk is a striking 66 arcseconds across, so the amount of light from even this partially illuminated face is considerable. Greatest brilliancy occurs about 36 days before or after the actual time of inferior conjunction, when the disk is almost one quarter illuminated.

> **Project 2: The Crescent Venus**
>
> It is worth catching Venus as it emerges from inferior conjunction simply for the sheer joy of seeing this brilliant planet in crescent phase and showing a large disk! A small telescope or even a pair of high-power binoculars is sufficient to catch the spectacle, and it truly is one worth seeing. Simply wait until the planet emerges from inferior conjunction and line it up in your telescope. Then enjoy the spectacle. No great astronomical research involved here, but it is worthwhile to remember that there are spectacles in the sky that are truly beautiful and ready to be enjoyed just for their own sake!

Not a Good Place to Visit!

The present time is an exciting one astronomically speaking. Not only are satellite-born instruments monitoring the microwave background—the very echoes of Creation—but accurate measurements of the proper motion and the brightness of other stars are detecting the telltale signs of other worlds. Whether these betray their presence in the slight wiggles of their host star's motion or through the very slight, but regular, dimming of its light as they pass in front of its disk, these extrasolar planets are turning up in vast numbers and astronomers are getting closer and closer to finding worlds matching our own in distant planetary systems (Fig. 1.4).

Yet, in one sense we have known of "a planet like our own" since time immemorial. Venus is the nearest planet of all in the spatial sense as well as being almost a twin of Earth in terms of size

Fig. 1.4 Venus (Credit: NASA/Ricardo Nunes. Image processing by R. Nunes http://www.astrosurf.com/nunes)

and mass. Sure, it is very slightly smaller and not quite as massive or dense, but the differences are negligible on the planetary scale of things. It is just over 406 miles (650 km) smaller in diameter, 19.5 % less massive and about 6 % less dense than Earth. Rightly is it often called Earth's sister planet and as I write these words, it is far more Earthlike than anything yet found by *Kepler* and other extrasolar planet searches (although I expect that this will soon change, and may indeed *have* changed by the time you read these words). Surely, if being physically similar to our home planet is what we mean by "Earthlike", then we have such a planet on our doorstep and, because the nearest of all planets is "another Earth" we can presume that the Universe teems with clones of our home world. If the physical similarity to Earth of our closest planetary neighbor is also reflected in its climatic and biological similarity, then the consequences would truly be profound.

Not surprisingly, when little was known about the conditions at the surface of Venus, many thought (understandably enough) that it might well be another Earth in these senses as well. An inhabited Venus did not seem at all farfetched. In fact, it appeared

20 Weird Worlds

Fig. 1.5 Surface of Venus showing craters Saskia (*foreground*), Danilova (*left*) and Aglaonice (*right*). This is a computer generated view created by superimposing Magellan radar images in topography data (Credit: NASA)

quite logical. Why should the planet that was nearest to Earth both in terms of actual distance and in size and general physical properties be other than like our own in terms of surface conditions and biology? After all, is it not a popular assumption that when *Kepler* or some such extrasolar planet-hunting project finally turns up a (physically) Earthlike planet orbiting some other star, it will be like Earth in these respects too?

In the case of Venus however, this assumption could not have been more wrong! Observations from spacecraft and surface landers have painted a picture of a planet harsher than anything ever imagined by earlier generations of astronomers. Back in the mid 1950s, Bart and Pricilla Bok wrote in an elementary astronomy book that "Venus, so beautiful from afar, would be a dreary place to visit." They pictured it as a desert world forever shrouded by a canopy of perpetual dust haze. Their "dreary" portrait of the planet was one of the understatements of the century, albeit the best that could be ascertained at that time. But it seems very benign compared with the picture that emerged during the following decades of space exploration (Fig. 1.5).

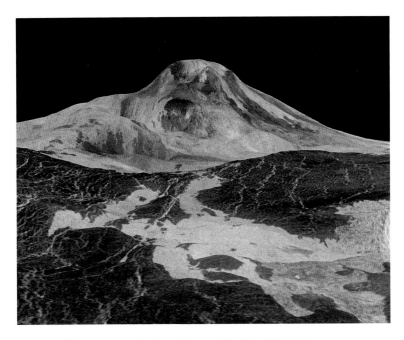

FIG. 1.6 Magellan image of Maat Mons (Credit: NASA)

The Boks were right about Venus being a desert, but the popular earlier picture of dunes and rocky pinnacles sculptured by wind-blown sand has been replaced by vast plains of basaltic rock broken into innumerable slabs. Some 80 % of the planet's surface is covered by these plains, divided between plains with wrinkle ridges (70 %) and smooth plains (10 %). The remaining 20 % of the surface is made up of two highland regions which have been described as "continents", albeit surrounded by oceans of rock, not water. We might imagine that if Venus was a watery world like Earth, 80 % would be ocean and 20 % continental land (Fig. 1.6).

One of these "continents" is in the northern hemisphere and the other in the southern. The former is known as *Ishtar Terra* and is roughly the size of Australia. It is the smaller of the two but has the distinction of being home to the planet's highest mountain, *Maxwell Montes*, which towers nearly 7 miles (11 km) above the average surface elevation, the Venusian counterpart of sea level. The Great South Land of Venus is known as *Aphrodite Terra* and is approximately the size of South America. Much of its area is covered by a network of faults and fractures.

The planet has about a thousand impact craters dotted evenly over its surface. This is a small number by planetary standards but they are all relatively large, ranging in diameter from just under 2 miles (3 km) to around 175 miles (280 km). Small craters are missing because smaller incoming bodies are slowed down by atmospheric drag to such an extent that they do not strike with explosive impacts and objects smaller than about 160 ft (50 m) in diameter break up before reaching ground level. Disturbed patches on the ground have been found and are thought to have been caused by the impact of blast waves of powerful airbursts from large exploding meteoroids (Venusian Tunguska's!).

Another feature of the impact craters of Venus is their comparative youth. Unlike the old weathered craters of Earth and the oldest Lunar ones which have been slowly degraded by ages of meteorite impacts, 85 % of those on Venus are in a pristine state, almost as if they were formed yesterday. We will come back to the reason for this shortly.

First however, a word should be said about two other features of the surface of this planet. Volcanic features are evident on Venus and, indeed, volcanism appears to have been the main force shaping the planet's surface. There are several times the number of volcanoes on Venus than on Earth and 167 of these are over 60 miles (100 km) across; larger than Earth's greatest volcanic formation; the Big Island of Hawaii. But in addition to volcanic mountains, craters and calderas, Venus has some volcanic features not found on any other planet in the Solar System. These include *farra*; flat-topped features looking somewhat like pancakes ranging in size from 12 to 30 miles (20–50 km) across and 320–3,200 ft (100–1,000 m) high, *novae* or radial star-like features, *arachnoids* (suitably named features having both concentric and radial fractures closely resembling spider's webs) and circular rings of fractures, sometimes surrounded by a depression, known as *coronae*.

Now, returning to the point raised earlier about the pristine state of Venusian impact craters, we note that part of the reason for the craters' relative youth and the reason for the large number of preserved volcanic features is the same; the planet's surface is considerably older than Earth's and is not experiencing the continual recycling via subduction that keeps our world's face fresh. Surface features, whether volcanic or meteoritic, stay pristine longer on Venus than on Earth … so long as they were formed during the last 300 million years. That is the other side of the story—*very* old features have *not* been preserved. But why? If conditions on the planet are such that relatively recent features are preserved better than their counterparts on Earth, why do the far older features that *have* been preserved on Earth (albeit in very weathered and distorted form) not have their counterparts (and better preserved counterparts at that!) on Venus?

The reason is believed to be due to a period of excessive activity on Venus between 300 and 600 million years ago. Thanks to its thick crust and relative lack of water, Venus cannot sustain the continual plate tectonics that occur on Earth. Without this "engine" to dissipate heat from the planet's mantle, the temperature of the mantle rises over time until a critical level is reached where the overlying crust is weakened to such an extent that a period of subduction takes place on a timescale of around 100 million years. This process is periodic; bouts of subduction taking place on an enormous scale during which the entire surface is recycled and all previous surface features erased. During these periods, mantle heat is dispersed, the mantle cools, the crust once more hardens and the planet settles down to another period of relative quiet. This episodic dispersion of internal heat, contrasted with the steady release experienced by Earth, is an important difference between the two planets and one that has far reaching ramifications for conditions on the two worlds. Not only does the steady dispersal of heat which Earth has managed to accomplish keep our planet from the violent periods experienced by Venus, but it is thought that the reduced heat loss of the latter planet is also responsible for another fact about Venus; its lack of an internally generated global magnetic field. As we shall see below, there is very little water on Venus today, although it was undoubtedly delivered to this planet by infalling comets and meteorites just as it was on Earth. But because Venus had no magnetic field, there was nothing to deflect the constant solar wind and molecules of water were quickly split into their components of hydrogen and oxygen in a way largely avoided on Earth. Quite rapidly on the cosmic timescale, Venus dried out. Thus, for more reasons than one, it seems that if Earth dealt with its mantle heat in the same way, life (at least at any advanced level) would have been impossible here and the two planets would indeed have been much more alike; though not in a way favorable to us!

During the times of recycling, Venus becomes a nightmare world of belching volcanoes and floods of molten lava. But does active volcanism completely disappear during the "quiet" periods in between as had initially been assumed? There is evidence that it does not.

One line of evidence comes from the existence of lightning in the planet's atmosphere. This was detected by *Venera 11* and *12* during the Soviet Union's *Venera* program. As well as recording a constant stream of lightning on its way down, a powerful clap of thunder was apparently recorded by *Venera 12* shortly after it landed on the surface of Venus. Not everyone accepted the *Venera* evidence, however the clear signature of lightning high in the planet's atmosphere was later detected by the European *Venus Express Mission* and its occurrence on Venus is no longer considered controversial. Whilst most lightning on Earth is driven by rainfall and ice crystals, neither of which occur on Venus (excepting sulfuric acid rain at high altitudes) it should be noted that not all terrestrial lightning involves these mechanisms. There are "dirty thunderstorms"—lightning generated in clouds of volcanic ash—as well as electrical activity sometimes reported in sand and dust storms. The suggestion has been made that the lightning on Venus is volcanic in origin and as such stands as evidence of continuing activity on that planet.

A second line of evidence concerns an otherwise mysterious drop in the sulfur dioxide levels in the Venusian atmosphere between 1978 and 1986. The concentrations fell by a factor of ten which is not easily explained unless the earlier reading was anomalously high, presumably the result of a massive injection of SO_2 into the atmosphere not long before 1978. A major volcanic eruption at that time would provide a very straightforward explanation for this.

Possible evidence was again noted in 2009 when a bright cloudy spot appeared on the planet and was observed from both Earth and space. The cause of this bright cloud remains unknown, but a volcanic eruption has been suggested by way of explanation.

Data from *Venus Express*, published in 2010, also provides strong evidence of geologically young lava flows and thermal imaging instruments on board the mission clearly detected warmth from a volcanic peak. This is good evidence that some of the planet's volcanoes were active in geologically recent times, although the findings do not necessarily mean that they remain active today. The *Venus Express* analysts estimate that the "warm" volcanoes were certainly erupting within the last 2.7 million and probably within the past quarter million years, "yesterday" in geological terms and certainly long since the last epoch of global recycling. Although contemporary volcanism is not proven by these results, at least they show that Venus does not shut down volcanically between the periods of intense global activity.

More controversially, data from NASA's *Magellan* space mission, also made public in 2010, has been interpreted by some scientists as indicating that one very fresh-looking and warm lava flow is probably no older than a few decades. This feature was first imaged in 1978, so was obviously formed prior to that year, but the proponents of its extreme youthfulness suggest that it may not be very much older. This has been challenged, or at least seriously questioned, by other scientists but if this formation really does turn out to be just decades old, is it possible that both it and the excess concentration of SO_2 found in 1978 tell of a powerful volcanic eruption on Venus, possibly as recently as the 1960s or early 1970s? If this does prove true, some scientists think that we may need to modify our view of intense volcanic periods separated by times of quiet. Venus may continually bubble along to a far more active degree than hitherto believed. A controversial speculation to be sure, but not one that is inherently unrealistic.

A World of Very Thick Air

In terms of surface conditions and global climate, the biggest difference between Venus and Earth lies with the great dissimilarity of the atmospheres of these worlds. Strong evidence for an atmosphere was provided by Mikhail Lomonosov's telescopic observations of the Venus transit of 1761, but ever since the first telescope users pointed their instruments toward the planet, the opaque nature of its atmosphere thwarted observations of its surface. True, not all astronomers blamed the bland appearance of its disk on a cloudy mantle but that came to be the majority opinion and must surely have been a frustration to the early telescopic observers of this world. Surely these pioneers entertained high hopes of seeing some really interesting features on this brilliant planet, but instead were confronted by a bland, if brilliant, disk going through its various phases.

Early speculations as to the nature of this cloudy covering ranged from water droplets similar to the majority of Earth's clouds, ice crystals, wind-blown dust, hydrocarbons and (once the signature of carbon dioxide was found in the planet's spectrum) polymers formed by the action of solar ultraviolet radiation on the atmosphere's most common gas. Depending upon one's opinion as to the nature of the clouds, the obscured surface of the planet was variously imagined as being a global ocean (of carbonic acid—soda water—once the presence of large amounts of atmospheric carbon dioxide were confirmed), a global tropical swamp (in some versions, complete with palms, jungles and saurian inhabitants), a planetary oil field, a hot desert shrouded in watery clouds, ditto but shrouded by high layers of icy clouds or a global dust bowl like the "dreary place" envisioned by Bart and Pricilla Bok. With little known about surface temperatures, Venus was thought by some astronomers to hold out the promise of life. A swampy or oceanic planet looked promising in this respect and English astronomer Patrick Moore was amongst those who speculated that the "soda-water seas" of Venus (if such did indeed exist) might harbor at least primitive aquatic life forms. A few brave souls even went so far as to suggest the possibility of *intelligent* life on the planet, although because these hypothetical beings could never see a clear night sky they would have no astronomical knowledge, no sense of their place in the wider Universe and must therefore still be barbarians!

Most of these speculations seem rather quaint nowadays in the harsh light of what space probes have revealed about our neighboring world. The atmosphere of Venus turned out to be a lot denser than had previously been imagined. Air pressure at the surface of the planet is a crushing 92 times greater than that at Earth's sea level. This is the equivalent of pressure some 0.6 miles (1 km) below the ocean surface. As suspected, carbon dioxide comprises the lion's share of the atmosphere. The pressure of this gas at the surface of Venus is so high that, technically speaking, it is no longer a gas but has assumed the state known as a supercritical fluid. Even a slight wind (perhaps we should say "flow"?)

is enough to shift small stones across the planet's surface. Walking through this dense fluid would also be very difficult but, as we shall soon see, that would be the least of one's problems if we could imagine something as bizarre as actually landing on Venus. Carbon dioxide is not, however, the only gas in Venus' atmosphere, although it does comprise 96.5 % of the Venusian air. Most of the rest consists of nitrogen with mere traces of sulfur dioxide, argon, water vapor, carbon monoxide, helium, neon, hydrogen chloride, and hydrogen fluoride. Hydrogen sulfide and carbonyl sulfide have also been detected high in the atmosphere and the early *Venera* space probes reported rather large amounts of chlorine just below the cloud deck. And those infamous clouds? They are for the most part composed of sulfur dioxide and droplets of concentrated sulfuric acid, otherwise known as oil of vitriol. Charming! Sulfuric acid rain falls from the clouds, but does not reach the surface. As it descends into the lower atmosphere, it evaporates as temperatures rise, wafting upwards to condense again at the cooler higher altitudes.

The large amount of carbon dioxide induces an horrendous greenhouse effect. Even though Venus receives 75 % less solar energy than Mercury (because of its greater distance), its surface temperature is higher; a blistering 860 °F or 460 °C! That is probably the best reason to doubt that humans will ever visit this world in person. Frying a robotic probe is much less controversial than risking a similar fate for a human being!

The surface of the planet retains a constant temperature, not even cooling down during the Venusian night (which, as we will see shortly, is remarkably long). Indeed, there is little difference in surface temperature between equator and poles, so we search in vain for a cool spot on the surface of this torrid world. There *is* a cooling of the atmosphere with altitude, so the tops of the tallest mountains are not quite so hellish. The summit of *Maxwell Montes* rises to levels where the atmosphere is close to 212 °F or 100 °C cooler than the surface. But that is still blistering hot! Ironically, the tops of some of the peaks appear to be covered in snow. The exact nature of this reflective substance is not known, but one thing is for sure. It is not water-ice snow! Most probably, it is some relatively volatile stuff that cannot condense on the surface but rises in gaseous form to the higher altitudes where it can condense as a snow-like covering on high mountains. Elemental tellurium or lead sulfide are amongst the suggested substances, but nobody is sure at this time.

There is something very weird about the way in which Venus' atmosphere behaves, but to appreciate this better, something equally odd about the planet's rotation must first be mentioned. Almost all of the Solar System's planets rotate about their axes in a counter-clockwise direction. This is why on Earth, for example, the Sun rises in the east and progresses across the sky to the western horizon. But Venus dares to be different. It rotates in a clockwise—technically called "retrograde"—direction. Moreover, it rotates so slowly that it takes the equivalent of 243 Earth days to make just one full revolution. Because a Venusian year lasts the equivalent of 224.7 Earth days, on Venus a single "day" is longer than a "year"! At least, that is true of a Venusian *sidereal* day (the time between successive crossings of the meridian by a specific fixed star). But the retrograde rotation of the planet means that a *solar* day (the time between two successive risings of the Sun) is significantly shorter than one sidereal day. It is actually "just" 116.75 Earth days long—shorter than the 176 Earth days of Mercury's solar day—making a Venusian year about 1.92 Venusian solar days long. To an observer on Venus, the Sun (forget about it being obscured for the moment!) would rise in the west and trek eastward across the sky, finally sinking beneath the eastern horizon nearly 117 Earth days later. Whether this peculiar rotation period and direction resulted from a giant impact during the early life of Venus or is the result of tidal effects and planetary perturbations—or something else again—is not known at present.

But whatever caused the slow retrograde rotation of the planet, a very weird fact is to be noted about the rotation of its atmosphere. It *super-rotates*! Thus, while the orb itself takes a leisurely 117 Earth days to complete a single rotation, the air surrounding it takes a mere four on average, with the atmosphere at middle latitudes rotating faster than that at the equator! Why this is so, nobody really knows, but ever since this odd feature of the planet and its atmosphere became apparent, astronomers reasoned that the rotation poles should be marked by massive, perpetual and stationary, super-hurricanes. Indeed, images from *Pioneer Venus* in 1979 revealed just such a vortex at the planet's north pole and it was expected that a similar feature should occur at the south pole as well.

An odd thing about the north polar vortex (odd at least when compared with terrestrial hurricanes) was that it possessed two eyes; two centers of rotation connected by S-shaped cloud formations. Rotation was in the direction of the super-rotating atmosphere and wind speeds close to the outer rim were measured as being in the range of 79–112 miles per hour (126–180 km/h) and, just like terrestrial hurricanes, essentially calm within the vortex eyes.

When *Venus Express* arrived at the planet in April 2006, astronomers were not in the least bit surprised when it beamed back images of a south polar vortex in every way similar to its northern counterpart, right down to the double eye. This, it seemed, confirmed the suspicion that these features were truly permanent features of the planet's atmosphere. But an unexpected surprise was in the offing. Very soon after first imaged, the double-eyed vortex changed completely, morphing into a single eye. Apparently, the polar vortices of Venus can assume a number of forms; the double vortex simply being one of them. Co-incidentally, it was the one that both polar vortices had assumed at the time we first observed them. How easily we can form the wrong opinion when a small amount of data appears to present a consistency that disappears with further observations!

A further strange feature of the polar vortices as revealed by *Venus Express*—more specifically, by the *Virtis* (*Visible and InfraRed Thermal Imaging Spectrometer* on board the space probe—is that the polar vortex rotates almost as if it were a solid body. This is not how the rest of the planet's atmosphere behaves. On the contrary, wind speeds close to the equator vary steeply with increasing altitude, even doubling between the level of the lower clouds and the cloud tops. Between approximately 60° north or south of the equator, circulation is such that the Venusian atmosphere below about 30 miles (50 km) effectively overturns. This convective overturning—air rising at the equator and descending at high (around 60°) latitudes—is called *Hadley circulation*.

Between the 60° north and south limit of the planet-wide Hadley cell and the polar vortices—in a region ranging in latitude between 60° and 70°—there exist zones separating the rotating vortex regions from the rest of the planetary circulation.

These constitute *cold polar collars* or rings of cooler air. Clouds within the collars reach about 3 miles higher than those at the poles on one side and at middle latitudes on the other. They are also denser than those of surrounding regions. Truly, the planet's atmosphere is full of surprises!

Like Earth's envelope, the atmosphere of Venus is differentiated into several more or less distinct layers. Again like Earth's, most of Venus' atmosphere (99 % of its mass in fact) is concentrated within the lower level or *troposphere.* This region extends up to about 40 miles (65 km) from the surface, but it is the first 31.3 miles (50 km) that account for 90 % of the atmosphere's mass. By comparison, 90 % of the mass of Earth's air lies within just over 6 miles (10 km) of ground level.

A little above this level, conditions become more Earthlike than anywhere else in the Venusian atmosphere. This region is known as the *tropopause* and represents the boundary between the troposphere and the mesosphere, about which more will be said in a little while. Earth's atmosphere also has a tropopause, in this instance dividing the meteorologically active troposphere and the layered stratosphere at around 7 miles (11 km) above the surface; considerably lower than its Venusian counterpart. At the 31—mile level of the Venusian tropopause, air temperatures range from about 68 °F (20 °C) to around 99 °F or 37 °C. Just below the tropopause—at 31 miles (49.5 km) above the surface—the atmospheric pressure equals that at the surface of Earth.

Ascending further from the hot Venusian surface, the region of atmosphere immediately beyond the upper level of the tropopause is known as the mesosphere. It extends between 40 miles (65 km) and 75 miles or 120 km. It is divided into two layers, the lower mesosphere to about 46 miles (73 km) and the upper mesosphere extending out to 59 miles or 95 km. Beyond this again is the *thermosphere,* extending to somewhere between 138 and 219 miles (220–350 km) after which the atmosphere merges with outer space. This outermost fringe, where atmospheric particles are so few and far between that they do not collide with one another, is known as the *exosphere.*

A layer of sulfuric acid haze exists near and immediately beneath the tropopause and the upper cloud layer coincides with the lower mesosphere. At least, it coincides with this layer during daylight hours. At night, it extends as a thick haze almost to the top of the upper mesosphere and (in the form of a thin haze) to as high as 66 miles (105 km); not far from the lower reaches of the thermosphere.

Temperatures in the lower mesosphere remain pretty much constant at −45 °F (−43 C) but fall again through the upper mesosphere, reaching −162 °F (−108 °C) at 59 miles (95 km) where the *mesopause* (the boundary between the mesosphere and thermosphere) begins. That is the coldest region of the dayside atmosphere. Temperature in the dayside mesopause itself rises again to a constant between about 81 and 260 °F (27–127 °C) in the thermosphere. On the other hand, the nightside thermosphere or *cryosphere* is the coldest region within the entire atmosphere (or the entire planet in fact) where temperatures plunge to a very chilling −279 °F or −173 °C. In 2011, *Venus Express* discovered a thin ozone layer within the mesopause, at about 63 miles or 100 km altitude.

The thermosphere has been found to almost coincide with an extended ionosphere extending from about 75–188 miles (120–300 km) above the planet's surface. This is pretty much a daytime phenomenon only as levels of electron concentration drop to nearly zero on the nightside. The upper boundary or *ionopause* is located between 138 and 234 miles (220 and 375 km) altitude. This is where the ionospheric plasma meets that of the solar wind wrapping around the planet. Because Venus has no global magnetic field (probably due to its very slow rotation) it has only an induced magnetosphere created as the lines of magnetic field carried by the solar wind wrap around the planet in the way water in a fast flowing stream "wraps around" an intruding rock. The induced magnetosphere creates a bow shock and a *magnetotail* as well as a *magnetopause* about 31 miles (50 km) above the ionopause. Between these two, the magnetic field is enhanced, creating a barrier against deeper penetration into the planet's atmosphere by the solar wind. At least, that was the situation when this phenomenon was observed in 2007 when the Sun was rather quiet. What happens when our star becomes stormy again remains to be seen at the time of this writing. In any case, Venus' lack of *intrinsic* magnetic field leaves it a lot more vulnerable than Earth to atmospheric erosion by the solar wind. This can penetrate well into its exosphere, sweeping ions away along the magnetotail (which, by the way, extends out to about ten planetary radii and forms the most active part of the magnetosphere). This stripping away of ions is thought to account for the dryness of Venus at the present time. As mentioned earlier, there is no reason to suppose that Venus missed out on a water supply similar to that of Earth's, but there is very, very, little of that remaining today.

The upper atmosphere of Venus experiences totally different circulation patterns from those of the lower. Between 56 and 94 miles (90 and 150 km) the atmosphere moves from the dayside to the nightside of the planet, rising over the dayside and sinking over the darkened hemisphere. Over the nighttime hemisphere, this sinking air warms and forms a warmer layer within the nocturnal mesosphere between altitudes of 56 and 75 miles (90 and 120 km). This layer is not all that warm—about $-45\ °F$ ($-43\ °C$) but compared with the typical temperature found in the cryosphere ($-279\ °F$ or $-173\ °C$) it is positively torrid! Interestingly, the air carried from the dayside brings with it atoms of oxygen which recombine on the night side to form excited oxygen molecules. As these relax from their excited state, infrared radiation at a wavelength of around $1.27\ \mu$ is emitted and has been observed both from Earth and from spacecraft.

Venusian Life?

At first sight, this section must seem like a waste of space. How could there possibly be life on a Hell-hole of a planet like Venus? Why should we spend time even thinking about such things?

Well, strange as it may seem, the question of Venusian life is far from being a simple open and shut case. Earlier we mentioned that Venus very likely received a similar amount of water as Earth but subsequently lost nearly all of this early water through the action of solar wind impinging upon its atmosphere; a process which was ultimately allowed to continue due to the planet's lack of intrinsic magnetic field. The reason for thinking that Venus was originally a wet planet comes mainly from the duel facts of its being composed of the same type of material as Earth and exposed to the same barrage of comets and meteorites as our home planet. If water outgased from early Earth, there is no valid reason to think that it did not from the early Venus as well. If Earth picked up most of its water from colliding comets and water-bearing meteorites, there is every reason to believe that similar projectiles pummeled our sister planet in the Solar System's youth, delivering water there as well.

Around four billion years ago, it is probable that Venus, Earth and also Mars, resembled each other far more closely than they do today. Indeed, they probably resembled each other more closely then, than any of the three resembled what they are today; rather as three 6-month-old babies resemble each other more than they do their mature selves at, say, 60 years of age. As we mature, life and circumstances determine much of our nature and to a certain extent, even our appearance. The same is true of planets. Earth, Venus and Mars are each thought to have possessed liquid water and relatively benign conditions toward the end of what on Earth is known as the Hadean era. Although named after Hell, later Hadean times are now believed to have been a good deal less severe than was once thought. We will return to this point shortly.

The oldest terrestrial rocks date back to around 3.7–3.8 billion years and are found to contain microscopic structures that look somewhat akin to microfossils of early bacteria found in somewhat more recent rocks. The biological interpretation of these earliest structures is controversial, but if the scientists who go with this interpretation are correct, it means that relatively complex micro-organisms were around at the time of the oldest surviving rocks. The oldest rocks are metamorphic; sediments from even older rocks that had been subjected to heat and pressure since first being laid down. Some of the particles from which they were initially formed date back to Hadean times and examination of very ancient rocks found in Western Australia uncovered the presence of zircon crystals dating back at least 4.2 billion years—far older than the oldest surviving rocks themselves and going right back into the Hadean past. Analysis of these crystals revealed some surprising facts about the environment in which they were formed. It was not very hellish at all. Indeed, it was the sort of environment in which liquid water—and maybe early life—could exist. In other words, if the microstructures found in the oldest surviving rocks really are micro-fossils, they might be the descendents of simpler organisms that were around in the latter ages of the Hadean era.

It is but a short step to postulating that if Venus and Mars harbored similar conditions at that time, simple life forms may have appeared there as well.

But, even if these speculations are correct (and we must never forget that they *are* just speculations!) can we reasonably think that life continues on Earth's neighboring planets today?

Before attempting to answer this question, let's imagine that the Solar System is being examined through a super telescope by an alien on a planet of another sun. Analyzing the atmospheres of the Solar planets, our hypothetical alien would almost certainly conclude that Earth was the most likely place to find life in this System. Why? Simply because it's atmosphere is so far out of chemical equilibrium. It has, as a major atmospheric constituent, a highly reactive gas (oxygen) that could not possibly be retained in such quantities over time unless it was being constantly replaced by a dynamic process. Life, the alien would almost certainly conclude, is the most likely process in this instance. This conclusion does not depend upon the alien being an oxygen breather himself. The conclusion would appear equally logical to a chlorine breather. The life signature is not oxygen *qua* oxygen, but the massive departure from equilibrium that large amounts of this gas implies.

Coming back to Venus (we will leave Mars aside for now), a broad overview of its atmosphere shows no such gross departure from chemical equilibrium. This appears to tell against contemporary life. However, the situation is not quite as straightforward as this. Lack of gross departure from equilibrium only implies lack of the sort of profound biosphere that exists on Earth.

A closer examination of the higher atmosphere of Venus does indeed reveal a lack of chemical equilibrium, albeit one more subtle than found on Earth. We mentioned earlier that *Venera* found significant amounts of chlorine beneath the cloud deck. Yet chlorine is a very reactive gas, so how is it maintained and where did it come from in the first place? *Something* must be continually producing it. This is not necessarily life, but some reaction must be progressing high in the Venusian atmosphere.

At even higher altitude, hydrogen sulfide and sulfur dioxide have been found together. These two gases react with each other, so the presence of both of them means that the atmosphere is out of equilibrium, implying the presence of some process (probably in the upper atmosphere itself) continually generating them. Additionally, carbonyl sulfide has been detected at these altitudes. This is a gas that is not easily produced by purely inorganic means.

On Earth, life has spread to just about every nook and cranny of the planet—surface, oceans, atmosphere and even deep within the crust. From this single example, we might suppose that once life gets a toehold on any world, it will eventually conquer that world. Moreover, life has changed the environment of our planet in such a way that our world has become even more life friendly. The appearance of oxygen-producing micro-organisms prepared the way for more complex oxygen-breathing life forms, which in turn led to the further spread of life. The Gaia hypothesis, if taken to the ultimate, seems to imply that life will eventually change its environment for its own benefit and a strong adherent of this philosophy might argue that something like the small disequilibrium of Venus' atmosphere is not nearly so great as to be indicative of life's presence.

However, the Gaia hypothesis is far from being the last word (see my *Weird Weather* for more about the various factors involved in the evolution of early Earth). Rather, it seems that the physical evolution of Earth and the growth of its biosphere worked together to effectively open up the planet to become one great ecological niche. Venus (and Mars too, we might mention) went the other way. If life ever got a toehold there, subsequent planetary evolution worked to restrict, not expand, its potential environment. The surface of Venus became an oven. Any liquid water disappeared. The surface periodically recycled. Life—if it ever existed—could only survive if it retreated to the one relatively benign environment on the entire planet; the upper atmosphere. Exactly where, please note, the interesting disequilibrium is found!

If life is ever positively identified in the atmosphere of Venus, it will show that the sort of co-operation between life and planetary evolution that has led to Earth's being completely covered by living organisms is not necessarily common to all planetary biospheres. "Restricted biospheres" like the one proposed for Venus may be more common in the Universe than "global biospheres" like Earth's, but they will also be more difficult to detect by reason of their more subtle nature. In any case, the indisputable discovery of such a biosphere on Venus would show that such "restricted biospheres" do exist and that life, once it finds a toehold on a planet, does not necessarily colonize every region of that planet.

But before we get too philosophical about the significance of Venusian life, we must ask whether this speculation rests on anything more than a relatively minor departure from equilibrium of part of the planet's atmosphere. The answer is actually a cautious "yes". In addition to sulfuric acid droplets, solid particles of unknown composition have been detected in the cloud deck. What these particles are has not been determined, but from the way in which they reflect light, they appear to have a non-spherical shape and a size consistent on both counts with bacteria. That does not necessarily mean that they *are* bacteria, but it does mean that they *might be*, other things being equal. In fact, sulfur metabolizing bacteria could thrive happily amongst the clouds of Venus. The greatest hazard is the strong ultraviolet radiation from the Sun, but even that could be rendered harmless if the microorganisms were coated with a layer of S8. This acts as a very efficient sunscreen and could quite adequately protect Venusian organisms against damage by this radiation. It might even by that these hypothetical microorganisms are adapted to harmlessly absorb solar UV and use it as a principal source of energy. Ultraviolet images of the planet reveal the existence of mysterious dark patches for which no adequate explanation has yet been forthcoming. The suggestion has been put forward that these might be dense floating colonies of UV-absorbing organisms, showing dark in images at that wavelength because the radiation is being absorbed by their bodies and not reflected back into space.

It goes without saying that none of this proves the existence of aerial life on Venus, but it does give enough circumstantial evidence to keep debate on the question open. Maybe a floating probe placed in the planet's high atmosphere will one day collect samples of Venus' air, and whatever else might be present there, and settle the issue one way or another. Indeed, because of its relative proximity to Earth and the high probability that any life that may exist there will inhabit the high atmosphere rather than the surface or subterranean niches, Venus should be considered the easiest major planet to test for the presence of life. It would be ironic if the first sure evidence of extraterrestrial life comes from this weird hellhole of a world, but that may well turn out to be the situation!

Is There Anything in a Name?

Before leaving this harsh but interesting world, it is worth noting that, like Mercury, Venus was given separate names for its morning and evening appearances. To the Greeks, the planet was known as *Phosphorus* in the morning and *Hesperus* in the evening. To the Romans, the equivalent names were *Lucifer* and *Vesper*. But once again like Mercury, it is hard to believe that amongst the early Greeks and Romans nobody tweaked to the fact that the two were identical. This is even harder to credit than the corresponding case of Mercury as both Phosphorus/Lucifer and Hesperus/Vesper vie with one another as the brightest regular denizens of the sky, excepting only the Sun and Moon. It is even stranger in view of evidence that the Babylonians apparently understood the two to be a single object at least as far back as 1581 BC. What is now known as the *Venus tablet of Ammisaduqa*, dating from that time, describes the planet as the "bright queen of the sky" and presents observations supporting the view that it appears alternatively both as a morning and as an evening object.

Not surprisingly in view of its brilliance, Venus is often mentioned in ancient writings and legends. As the morning star, it is mentioned in the Bible, where it is given the name *Helel*; a Hebrew word meaning "shining one" and approximately equivalent to the Latin *Lucifer* or "light bringer". In the Old Testament book of Isaiah (Chap. 14: 4–23), the prophet compares the tyrannical king of Babylon with the bright morning star, implying that the king shines in the political firmament as brightly as Helel shines in the celestial. But all that was about to change and this tyrant—the "morning star" of the political realm—was about to fall from his position of splendor. The inference is that, like the occasional politician before and after that time, this king had grown too big for his boots and was about to topple into oblivion. It has been suggested that Isaiah may have had in mind an old tale about the morning star trying to outreach its dominion until being overpowered by the rising Sun. In any case, his reference to Helel (translated as "Lucifer" in the Latin Bible) was directed at a very human tyrant, not a fallen archangel. This latter misreading of the passage came mainly with the Church father Origen and his penchant for finding symbols and allegories in scripture, but it has unfortunately entered into mainstream thought and this name for Venus has acquired a literally diabolical connotation, even though the only other biblical use of the name (in 2 Peter 1:19) has no connection with the devil—quite the opposite in fact!

Oddly, just about every name given to Venus has acquired a sinister connotation, at least in certain usages. Even Venus herself, though the goddess of love and beauty, was not (if I remember my mythology correctly) the nicest member of the Pantheon, but the alternative names for the planet can raise even less attractive associations.

Thus *Phosphorus* is also the name of an element which (though necessary for life) is extremely inflammable and has been used in incendiary devices in wartime. In the form known as either "white" or "yellow" phosphorus it is so inflammable that it must be stored under water. If exposed to air, it slowly "burns", giving off the pale glow for which it is widely known and can even be ignited by the warmth of a human hand ... with serious consequences for the handler!

Likewise, the name *Hesperus* brings to mind Longfellow's *The Wreck of the Hesperus*; a poem about the tragic consequences of a sea captain's pride. "Looking like the wreck of the Hesperus" is a phrase sometimes used to describe a place or person of badly disheveled appearance.

Only *Vesper* might appear benign; "Vespers" being the name given to the church service of evening prayer. But let's not forget the "Sicillian Vespers", the bloody uprising by the Sicillian population against the rule of the French king Charles 1, beginning at the hour of Vespers on Easter Tuesday in the year 1282 and resulting, over the following 6 weeks, in the wholesale slaughter of some 3,000 French men and women. The revolution was successful in so far as it freed the island from French rule, but only at the cost of triggering a conflict known to historians as the "War of the Sicillian Vesprers" which dragged on for the following 20 years!

Ironically, just as the planet itself has a duel personality—beautiful from afar, but something else again in its true nature—the names given to it have come to display a similar duality. Who says that there is nothing in a name?!

Mars

The next rocky planet outward from the Sun is a really weird one—Earth! At least, that is how it looks when compared with its neighbors. After all, it is teeming with that strange phenomena called life and, not only that, life that is manifested in highly complex forms. One of these forms is very odd and goes about building telescopes and studying the Universe! Yet, because this book is unashamedly Earth-chauvinistic, we shall overlook the familiar weirdness of our home world and proceed straight to the fourth rock from the Sun and the first one that lies beyond the orbit of Earth; the Red Planet, Mars.

A Controversial Planet

Mars has long been a planet of speculation and controversy. The latter decades of the nineteenth century were marked by a conviction that this was a smaller version of Earth, similarly inhabited by a race of sentient beings who were probably older and therefore a good deal smarter than us. Speculation was given a great boost in 1877 when Italian astronomer G. V. Schiaparelli produced one of the most detailed charts of Mars to that time and noted, in addition to the already well-known bright and dark regions of the planet, an apparent network of very fine and straight lines which he called "canali". That word is best translated into English as "channels", but unfortunately the most popular translation was "canals", implying artificial construction in a way which the neutral "channels" does not.

We all know about the hypothesis of Percival Lowell; intelligent inhabitants of a drying and dying planet constructing canals to bring precious water from polar regions to irrigate the inhabited but increasingly dry regions at lower latitudes. Given this "evidence" for intelligent Martian life, all manner of minor changes on the surface of the planet were interpreted as further indications of Martians. Transient spots of light (probably high clouds catching the Sun's rays) were interpreted as "signals", as were apparent patterns of marks on the disk of the planet and the famous W-shaped cloud that periodically forms over the region known as

Tharsis. Significantly however, Lowell himself paid little heed to these claims of Martian signaling. Although he thought that some of the transient features may relate to life in the wider sense (for instance, he explained what appeared to be a glowing spot as a possible fire in a Martian forest) he clearly scoffed at the suggestion that bright pinpoints of light seen on the surface were beacons set to draw the attention of Earthlings. On the contrary, he suggested that they were more likely clouds catching the rays of sunlight; an explanation revived again by Patrick Moore to explain a transient point of light observed by Tsuneo Sahecki many years later in 1951—and "explained" by some at that time as the explosion of a Martian atomic bomb!

The sobering images transmitted back by Mariner 4 in 1965 put an end to any lingering thought of intelligent Martians in the minds of most people, although the debate about life in some lowly form still continues, as we shall see in a little while. But for most of the general public, Mars has lost much of its former perceived weirdness. It is no longer the home of canal building beings, no longer the place where flying saucers are launched toward Earth, not even the home of the vast lichen prairies that were supposed to comprise the planet's dark regions that at an earlier time were mistaken for seas and other bodies of water. The light and dark areas are just regions that have rocks of a different hue. Nothing stranger than that!

Some of the issues and alleged evidence, past and recent, as well as the odd stories (also recent as well as ancient) that have been proposed in connection with Martian life were discussed in my *Weird Astronomy* and will not be repeated here, although some more recent contributions to the discussion will be raised in due course. For the moment, let's just look at Mars itself which—with or without life—is a very interesting world in its own right.

Mars is a small world; larger than Mercury but considerably smaller than either Earth or Venus, with an equatorial radius of 2,122 miles (3,396 km) or just over half that of Earth. It is also a good deal less dense than the other three rocks near the Sun, with a mean density of just over 3.9 times that of water. In terms of orbital eccentricity, Mars' 0.093 also comes in between the eccentric Mercury (0.2) and the more sedate Venus (0.006) and Earth (0.016). At its furthest point from the Sun (aphelion), Mars is 1.665861 AU distant (almost 156 million miles or 249 million kilometers) and at its perihelion or closest approach to the Sun, 1.381497 AU (approximately 142.5 million miles or 227.9 million kilometers). From our viewpoint in the Solar System, this rather eccentric orbit means that the planet can vary greatly in its apparent size and brightness; the maximum range covering the times when Mars is at aphelion and near conjunction with the Sun and when it is at perihelion and opposition (i.e. directly opposite the Sun as observed from Earth) at the same time as Earth's own aphelion passage. Between these extremes, Mars varies in apparent magnitude from 1.6 to −3 and its diameter from a mere 3.5 seconds of arc to a respectable 25.1 seconds of arc. Somewhat annoyingly for visual telescopic observers, the extra heating received by the planet around the time of perihelion makes this prime Martian dust storm season, so the otherwise very favorable perihelic oppositions, when one might expect the best views of the Martian surface, frequently end up yielding splendid views of a global dust storm that renders the Martian disk almost as bland as that of its big sister, Venus. These great dust storms dwarf anything seen on Earth. The entire atmosphere of the planet fills with dust, obscuring all surface features from the prying eyes of terrestrial astronomers. One such storm was in progress when *Mariner 9* achieved Martian orbit in 1972. Had this been a simple flyby probe, little would have been achieved. As it happened, the first surface features to peep through were the summits of the high volcanoes, especially *Olympus Mons*, offering a tantalizing hint that the planet was a far more interesting place than the earlier probes had suggested. One wonders what might have happened had *Mariner 9* not been an orbiter. Would Mars exploration have ground to a halt had it only recorded the dust storm and missed the volcanoes? (Fig. 1.7)

Dust storms notwithstanding, perihelic oppositions are interesting because at these times Mars approaches Earth more closely than any other major planet excepting Venus, reaching a minimum distance of just 35 million miles or 56 million kilometers from our world.

In common with Venus and Earth, but unlike Mercury, Mars is surrounded by an atmosphere worthy of the title, albeit far more tenuous than that of either of its two nearest neighbors. At the Martian surface, air pressure is a meager 0.6 % that of the sea level pressure of Earth. This is equivalent to the pressure of our

FIG. 1.7 Comparative sizes of Mars and Earth (Credit: NASA)

atmosphere at an altitude of 22 miles (35 km), significantly thinner than had been believed prior to *Mariner 4*. There is evidence that the atmosphere was denser in the distant past but because Mars does not possess a global magnetic field, much of its air has been blown away by the relentless stream of solar wind. Nevertheless, although the atmosphere is thin by our standards, it is still dense enough to give rise to a variety of weather phenomena. Dust storms and clouds have already been mentioned in passing and images from the surface of the planet have revealed high cirrus wisps not unlike those with which we are so familiar here on Earth. Whirlwinds are also frequently observed, both from the surface and from orbit. Some of these are truly gargantuan; one of them was observed to reach the incredible height of 12 miles (19 km)! Frost and fog has also been observed at the surface, but rain, hail and snow are not part of the Martian weather forecast. Four billion years ago when the atmosphere was thick enough to allow bodies of liquid water to exist, precipitation in these forms presumably occurred, but the thin atmosphere of today no longer allows it. Lightning and thunder are also absent, although it would be interesting to know if some milder form of atmospheric electrical activity might accompany some of the dust storms and dust devils that sweep across the planet. On Earth, strong electric fields have been found under such conditions and lightning is reported from time to time in dust/sand storms and even large dust devils. Could the same happen on Mars? Some astronomers think that in the rarefied Martian atmosphere, electrical discharges similar to the high altitude "sprites" observed high above terrestrial thunderstorms might accompany at least some of the dust devils that continually march across the surface of the planet.

In common with that of Venus, the air of Mars consists primarily of carbon dioxide; 95.32 % of the atmosphere being made up of this gas. In Mars' instance however, atmospheric density is not high enough to significant greenhouse warming. As one astronomer put it, Mars has a "leaky greenhouse". In addition to CO_2, the Martian air also contains nitrogen (2.7 %), argon (1.6 %), oxygen (0.13 %), carbon monoxide (0.08 %) as well as traces of water vapor, nitric oxide, neon, krypton, molecular hydrogen, deuterium oxide, formaldehyde, hydrogen peroxide and methane. The presence of the latter gas means that the atmosphere of Mars, once again like that of Venus, is slightly out of chemical equilibrium. Methane is quickly destroyed in the atmosphere and its continuing detection implies an active source of the gas. The nature of that source is hotly debated and we shall return to this in due course.

The surface of Mars is covered with rust. Or, to give it the more formal and polite title, hematite or iron (III) oxide. This is distributed over the primarily basaltic surface in the form of fine dust and it is this that gives the planet its characteristic red color, and indirectly its name. Mars is, of course, the name of the god of war and what better symbol of war than the color of blood—even if the "blood" of Mars is rather pale and the color of the planet somewhat anemic in comparison with a drop of the real stuff?

Project 3: The Color of Mars

Mars has long been popularly known as the "Red Planet" and when seen against a dark night sky, it is indeed contrasted beside the majority of stars by its distinctly ruddy hue. Yet just how red is it—really?

Continued

Project 3: (continued)

First, compare it with some bright "red" stars. Antares in Scorpius is a good comparison, as its very name means "rival of Mars". Betelgeuse in Orion is another. Does the color of Mars appear redder, less red or very similar to these?

Stars of this type (known as M-class) actually have temperatures quite similar to that of the filament of an incandescent electric bulb and radiate at about the same wavelength (= color). This is surprising, as we don't normally think of ordinary electric bulbs as being red! But if you have an opportunity to view one of the old-fashioned incandescent bulbs at a distance—such that its apparent brightness more or less matches that of Antares or Betelgeuse—you may like to compare their colors. Then compare the color with Mars. What is your conclusion about the color of Mars?

Try comparing Mars with the color of a distant stoplight, or some other equally red light source. How does the planet compare with this?

Dust storms on Mars were traditionally called "yellow clouds" by planetary astronomers. But they are composed of the same particles that give Mars its "red" color. "Yellow" might seem the wrong word to use for a planet named after the god of war but if the color of Mars is compared with, say, Antares or even a distant light globe at times when its atmosphere is free of dust and again during times of global dust storms (as frequently occur when the planet is near perihelion) does it show any discernible difference? If color wide-angle photographs of Mars in the sky (not images of the planet as observed through a telescope) taken during clear Martian atmosphere conditions on one hand and during a global dust storm on the other are compared, do they show any difference in color?

Internally, Mars possesses a partially-fluid core of about 1,121 miles (1,794 km) diameter, consisting mostly of iron and nickel but with a relatively large amount (16–17 %) of sulfur and around twice the concentration of the lighter elements than its terrestrial counterpart. This is surrounded by a mainly silicate mantle which in turn is overlain by a crust ranging in thickness between 31 miles (50 km) and 74 miles (125 km). By comparison, Earth's crust averages just 25 miles (40 km) in thickness and, considering the difference in the diameters of the two planets, it is clear that the crust of Mars is far more substantial than that of our home world. The thick Martian crust, plus the lack of lubricating liquid water, go a long way to accounting for another feature of Mars; its lack of plate tectonics and of a global magnetic field. At least, the present lack of these features, as evidence of early magnetism and even polar reversals have been revealed through the analysis of magnetically sensitive minerals in the planet's surface. It seems that about four billion years ago, Mars possessed both plate tectonics and a global magnetic field, but the nature of the thick crust and a drying of the planet prevented these processes from continuing. Once the magnetic field faded, atmospheric stripping by the solar wind increased, the planet dried out further and in time assumed the general form that we find today.

For such a small planet, Mars boasts some enormous surface features. Thus, a light colored patch observed through Earth-bound telescopes and rather romantically named *Nix Olympica* or the snows of Olympus, was found by *Mariner 9* in 1972 to be an enormous extinct shield volcano, similar in general type to the Hawaiian island volcanoes—only far larger. Now known as *Olympus Mons* (Mount Olympus) it is recognized as the largest known mountain in the Solar System, towering nearly 17 miles (27 km) above base level. This makes it over three times higher than our own Mount Everest. The reason this volcano is so large is undoubtedly due, at least in part, to the lack of plate tectonics on Mars. Here on Earth, the crust drifts over volcanic "hot spots" and passes on its way before any singular volcano becomes too large. But thanks to the immobile crust of Mars, its volcanoes just keep getting bigger and bigger until the hot spot activating them eventually grows cold. (Fig. 1.8)

Then there is the "canal" *Agathadaemon* or, to give it its post-Mariner title *Valles Marineris*. This is a stupendous canyon, around 2,500 miles (4,000 km) long and 4.4 miles (7 km) deep. Its length equals that of Europe and beside it, the Grand Canyon is little more than a ditch—a paltry 279 miles (446 km) long and 1.25 miles (2 km) deep. Not even the advanced Martians of Lowell's hypothesis could

FIG. 1.8 Mars showing Valles Marineris (Credit: NASA)

have engineered a "canal" of these dimensions! Planetary scientists now believe that *Valles Marineris* was formed by the swelling of the *Tharsis* region, home of *Olympus Mons* as well as several other large volcanoes. As *Tharsis* swelled, the region now occupied by *Valles Marineris* collapsed, forming this spectacular canyon. This is not the only colossal canyon on Mars, although it is by far the largest. Another sizable ditch is *Ma'adim Vallis*; over one and a half times the length of the Grand Canyon, of similar depth and around 12.5 miles (20 km) wide (Figs. 1.9 and 1.10).

What appear to be massive lava coils—similar in general appearance to those seen following eruptions of Hawaii's Mona Loa, except of far greater size—have been spotted in images from the *Mars Reconnaissance Orbiter*. Some of these are as much as 100 ft (30.5 m) across. This is from 50 to 100 times the diameter of their Hawaiian counterparts. The coils are thought to have formed as the surface of an ancient lava flow cooled and contracted faster than that lying beneath. The result is a splitting of the surface lava crust into a polygonal pattern. As the warmer underlying lava remained fluid for a longer time, it continues moving downstream in the process twisting the surface polygons into spiral shaped coils.

FIG. 1.9 Atmosphere and surface of Mars (Credit: NASA)

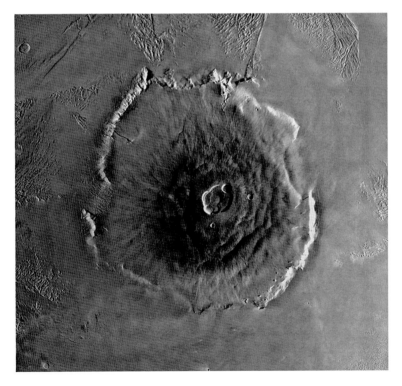

FIG. 1.10 Olympus Mons (Credit: NASA)

FIG. 1.11 Victoria crater (Credit: NASA)

The canyons, mountains, lava patterns and so forth were the result of processes native to Mars itself, but the planet is also strongly suspected of sporting an impact feature having an area of some 40 % that of the entire Martian surface. This huge northern hemisphere basin is theorized to have resulted from an ancient impact by a body about the size of Pluto!

Days, Nights, Seasons and Polar Ice

One of the features of Mars that convinced early telescopic observers of its Earthlike nature is the remarkable similarity between the length of the Martian and terrestrial day and the surprisingly Earthlike inclination of its rotational axis. One sidereal day on Mars is just 0.026 (terrestrial) days longer than a sidereal day on Earth; a remarkable coincidence. Similarly, the tilt of the planet's axis is 25.2°; remarkably close to Earth's 23.5°. This close similarity is just a passing phase however. While the presence of the Moon keeps the variations in Earth's axis quite small, Mars has no large companion world to keep it on the straight and narrow. Its axial tilt varies by over 20° from the present value so most of the time the striking similarity which we perceive today simply does not exist! The contemporary tilt means, however, that Mars currently experiences seasons similar to those of Earth, though because its "year" is nearly twice as long as that of our planet, the duration of its seasons vary accordingly (Fig. 1.11).

When Mars is viewed through even a small telescope at opposition, one of the features most readily seen is a polar ice cap; north or south depending upon the Martian seasons at the time of observing. Early telescope users found that the polar caps varied with the Martian seasons and (in keeping with the generally Earthlike picture of the planet so popular prior to the advent of space exploration) formed the opinion that they were comprised of thin layers of water ice that built up into frost-like layers during winter and evaporated with the somewhat warmer conditions of summer. Lowell and some of his contemporaries thought that liquid water flowed away from the melting caps (directed, in Lowell's opinion, into irrigation canals constructed by intelligent Martians) and, perhaps, revived fields of vegetation at lower latitudes as evidenced by reports of darkening and even changes of color in the darker areas of the planet. A popular (but never universally accepted) notion was that a "wave of darkening" spread from the shrinking ice cap toward the equator. Further observation modified this picture however. Even before the first space probes found the Martian atmosphere to be far thinner than anyone expected, it became known that Mars simply did not have enough air pressure to permit water to remain liquid on its surface for long enough to flow through channels. Whatever the "canals" were, they were not waterways; either artificial or natural. And if there is a wave of darkening and if it is due to vegetation responding to moisture, that moisture must come from increased atmospheric humidity, not flows of surface water. This humidity could not be great, it was concluded, as the degree to which the polar ice shrinks must mean that it is just a thin layer. Indeed, the south polar cap can even disappear altogether in the summer time. Mars began to look dryer than even Lowell had imagined it to be.

Some astronomers went so far as to question whether the polar caps were deposits of water ice at all. Frozen carbon dioxide looked like a viable alternative but in 1960 C.C. Kiess, H. K. Kiess and S. Karrer, proposed the even more radical thesis that the caps were actually deposits of solid nitrogen tetroxide! These astronomers presented a model whereby the atmosphere of Mars contained significant amounts of oxides of nitrogen and the color of its surface was evidence, not of oxides of iron, but of nitrogen peroxide. The Mars envisioned by these authors was a truly toxic planet, totally inhospitable to any imaginable form of life, and probably to any *unimaginable* form as well!

The Kiess/Karrer model won little acceptance at the time and has since been proved totally wrong. On the other hand, the hypothesis that frozen carbon dioxide ("dry ice" as it is popularly known) makes up the caps got a boost from the earlier *Mariner* spacecraft when this substance was indeed found at the poles. For many people, this appeared to mean a dry Mars and seemed to put an end to any remaining hope of finding life there. It may not have made the planet as hostile as Kiess, Kiess and Karrer thought, but it surely put an end to all talk of vegetation fields reviving as humid air drifted across them.

As not infrequently happens however, further research—in this instance coming mostly through subsequent space exploration—showed that the situation was not so much *either/or* as *both*! The white layers of dry ice formed—literally—only the tip of the iceberg. Beneath this seasonal cap of frozen CO_2 lay deposits of water ice, the extent of which not even Lowell could have dreamed. Mars is not deprived of water. There is (with apologies to S. T. Coleridge) "water, water everywhere" … though it is also true that there is "not a drop to drink". At least, not without melting it first. Far from the water supply of Mars being confined to frosty polar deposits scarcely large enough to fill a large lake (as even the staunchest "polar-caps-are-water-ice" proponents thought just prior to the first *Mariners*), we now know that if the ice at the south pole alone was melted, the entire planet would be flooded to a depth of about 36 ft or 11 m. But there's more! Ice is not confined to the poles. Quite the contrary, a zone of permafrost extends to around 30° of latitude from each pole. Mars is an icy, but certainly not a dry, world! (Figs. 1.12 and 1.13)

Like the poles of Earth, those of Mars experience chill darkness during the depths of winter. It is during these times that carbon dioxide freezes out in large quantities overlaying the permanent caps of water ice. As much as 25–30 % of the Martian atmosphere freezes out onto the winter pole. But come spring, the feeble warmth of the distant Sun is enough to evaporate much of this, creating a region of comparatively high pressure over the relevant pole. Seeking equilibrium of atmospheric pressure, this gas—freshly released from its frozen state—floods outward from the polar regions and toward the lower pressure environment at lower latitudes. As it flows outward, it creates powerful gales of 250 miles (400 km)/h, blowing before them clouds of dust and water vapor and triggering the formation of very Earthlike clouds of (water) ice crystals—cirrus clouds or "mares tails" as they are popularly called here on planet Earth.

These powerful winds blowing outward from the poles, not the water vapor carried along by them, are most likely responsible for the surface changes that earlier generations of astronomers

FIG. 1.12 Martian north polar ice cap (Credit: NASA)

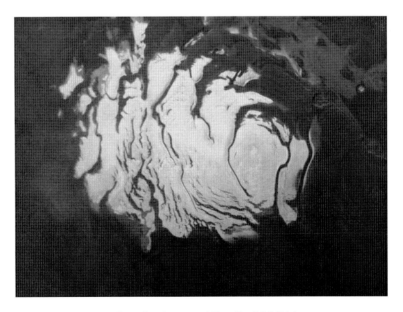

FIG. 1.13 Martian south polar ice cap (Credit: NASA)

mistook for fields of vegetation reviving under the onset of higher humidity. Dust that settled over dark rocks during the quieter months of winter gets blown away in spring, revealing more of the darker underlying surface. From Earth, the dark feature grows darker still and may even change color to some degree as the rocks are "cleaned" of lighter colored dust. An especially strong wind might even uncover a region of dark rocks that had been buried for years, causing the appearance of a fresh feature (a dark patch the size of the state of Texas appeared in 1954 and, true to the prevailing belief that these regions were vegetated, was suggested by at least one writer as possibly being the result of a Martian desert reclamation project! The real cause was considerably more mundane). On the other hand, wind-blown dunes might completely smother other outcrops of dark rock, resulting in the disappearance of certain features. Larger features remain pretty much permanent, although a widespread stripping of fine dust might cause them to darken somewhat. As the features closest to the pole are the first to be hit by the seasonal hurricane, any alteration will show up first in these and in this way the "wave of darkening" might find some grounding in fact, although it is suspected that the alleged "wave" is a lot less regular and continuous than described by the more enthusiastic supporters of the reviving vegetation hypothesis. When observing near the limits of visual discernment—which is the situation with Mars even at the best of times—one can easily (and quite subconsciously) make what is seen fit with what one would *like to see* or what one believes *should* be seen!

This seasonal accumulation of dry ice, at least at the south pole of the planet, gives rise to what must be one of the strangest sights in the inner Solar System. Images from the *Mars Global Surveyor* revealed the growth of mysterious spider-like or web-like patters of fine streaks radiating outward across the water-ice polar cap from central points, some of which coincided with craters. The propensity for these features to grow with the onset of spring led some scientists to suspect lowly forms of life as the cause. The answer, however, lies not with life but with something quite alien to our Earthly experience. The real culprit is the seasonal freezing out of carbon dioxide. During the dark and cold of winter, the seasonal CO_2 deposit starts out as a dusty layer of dry-ice frost covering the permanent water ice cap. As winter deepens, the frost layer re-crystallizes and grows denser, the dust particles caught in

Fig. 1.14 Artist's impression of erupting CO_2 jets on Mars (Credit: NASA)

the frost slowly sinking toward the bottom. By the winter's end, what started the season as a layer of frost has turned into a slab of semi-transparent dry ice about 3 ft (roughly 1 m) thick, lying over a substratum of dark dust. Being semi-transparent, spring sunlight passes through the slab and warms the bottom (dusty) layer, causing the dry ice there to sublimate. As sublimation continues, gas builds up beneath the transparent slab, finally lifting and rupturing it at weak spots and erupting through the breaches in dusty fountains of pressurized gas. Gas laden with dark basaltic dust rushes toward the vent, carving out groves which converge on the vent and give rise to the mysterious web-like patterns of streaks. Hopefully, tourists from Earth will one day send back holiday pictures of these strange gas-and-dust fountains (Fig. 1.14).

Martian Landslides and Gullies: Wet or Dry?

The spider webs of the Martian South Pole are not the only rapid changes that orbiting space probes have detected on Mars. Land slips and slides have also been found. The *Mars Global Surveyor* found that one such slide happened within the steep slopes of a crater sometime between the images of this region taken in August 1999 and September 2005. Other landslips imaged by this same craft the following year revealed flows of material that appeared to have weaved their way around obstacles and branched out into tributaries not unlike the pattern of a flash flood here on Earth. Moreover, these markings were not present in images taken just 4 years earlier and so constituted *very* recent events. According to some scientists, the most likely cause of these features is liquid water. And this explanation is thought by its supporters to explain more than simply the occasional landslide. Mars boasts hundreds of thousands of gullies that have the appearance of having been carved out by sudden flows of water—Martian flash floods if you like! But how could this be? How can a world with an atmosphere so thin that liquid water immediately boils at its surface maintain a surface flow long enough to carve out a gully or trigger a landslide?

The following scenario has been suggested. Groundwater migrates along rock layers until it becomes exposed at the surface, usually at the steep slope of a crater wall or valley. This causes a collapse at that point, with icy soil forming a plug preventing further percolation of groundwater toward the surface. Over time, water pressure builds up behind this ice dam, eventually breaching it and flooding down the crater slope in quantities too great for evaporation to immediately clear it away. This theory has some support. Gullies of this type are most often found on poleward-facing slopes at latitudes greater than 30°. (Beyond 30° south actually, as over 90 % occur in the southern hemisphere). Moreover, as already remarked, the form these gullies takes is very much as one would expect from flows of water. Furthermore, the one observed in 2006 had bright markings which could be interpreted as, either, fresh deposits of sediments left behind by the muddy flow, or as frost formed as water evaporated from the mud flow, struck the cold air and froze on the ground as ice.

The water ice explanation does, however, have its critics. For example, Serina Diniega and colleagues of the University of Arizona noted changes in gullies on the faces of sand dunes in seven locations south of −40° during the winter months when water would certainly be in solid form. The temperatures at ground level might actually be cold enough for carbon dioxide to freeze and these scientists suggested that dry ice frost may accumulate on a dune until it becomes thick enough to start an avalanche down the slope, dragging other material along with it as it goes. Perhaps significantly, the morphology of the gullies formed on the slopes of these dunes is very similar to the thousands of others noted on the planet, hinting that dry ice rather than liquid water might indeed be the real culprit in all of these.

A related process was suggested by Yolanda Cedillo-Flores and colleagues at Universidad Nacional Autonoma de Mexico and the Lunar and Planetary Institute USA. They proposed that dust and sand may lie on top of carbon dioxide ice in polar regions. As sunlight warms the ground in spring, frozen CO_2 sublimates and flows upward through the dusty layer above it. Experiments conducted by Cedillo-Flores show that if enough gas flows through a layer of dust or sand, that layer becomes fluid and literally flows down the slope almost as if it were liquid. A "dry" flow like this would be very difficult to distinguish from a "wet" one. It becomes virtually impossible to tell which Martian gullies are "dry" and which (if any!) are "wet" just by looking at them.

Such arguments have not converted the "wet" camp however. During 2011, the *Mars Reconnaissance Orbiter* found literally thousands of fingers of dark material appearing on steep slopes during the summer months. Unlike the gullies, these were in equatorial regions although the slopes themselves faced toward the middle latitudes. Carbon dioxide frost could not exist at these "warm" locations, but temperatures were nevertheless too low for water to be liquid. At least, for *pure* water to be liquid! Brine (salt water), by contrast, melts at a lower temperature than the pure liquid and it has been suggested that the dark streaks might be caused by flows of brine. It would need to be very salty—probably more like the Dead Sea than the blue Pacific—but that is not a problem in itself. On the other hand, there is absolutely no evidence at this time that water, salty or otherwise, is present in association with these streaks. Moreover, even if brine is positively identified as being the cause of *these* seasonal dark streaks at temperate latitudes, the scale of such structures and the general temperature environment of their locations differ so significantly from the types of gullies discussed above that it does not automatically follow that brine is the culprit there as well.

So the arguments continue without, at the time of this writing, any clear solution. All are agreed that water (as ice) exists on Mars and played its role in sculpturing the planet during those distant times when Mars was more Earthlike than it is today. Giant meteorite impacts must melt large volumes of sub-surface ice and gushers of liquid water following these impacts presumably account for the skirt-like features surrounding some impact sites. But whether the many recent gullies—including those seen forming today—are wet, dry or a mixture of both is still an open question, albeit one that should be solved in the not-too-distant future, given the continuation of Martian space research.

Whilst on the subject of landslides however, we should also draw attention to odd dark streaks seen near some very fresh impact craters to the south of the great *Olympus Mons* volcano. Sometime between May 2004 and February 2006 a number of meteorites struck this region of Mars, blasting out five small craters. The largest is around 72 ft (22 m) across. These were probably the work of a fragile body entering the planet's atmosphere and breaking up into a cluster of large lumps which then struck the surface as individual impactors. The dark streaks have been interpreted as being material exposed by avalanches triggered by shockwaves from the impacts. Mysterious parabolic dark marks also emanate from the center of the largest crater. Given the name of "scimitars" these features were apparently formed by the interference between pressure waves in the Martian atmosphere from both the incoming meteorite(s) and a hemispherical shock wave expanding away from the impact site and actually elevating the dust above the surface just prior to the impacts themselves. Alerted to this pattern, scientists have since found similar signs of avalanches and scimitars associated with other impact craters, thereby uncovering yet another "unearthly" feature of the Red Planet!

The Lure of Life

In the popular mind, even following decades of robotic exploration with no definitive results, mention of Mars raises thoughts of extraterrestrial life. Traditionally, this has been the planet thought most likely to harbor something which could dimly be called "life" and it is true to say that although these hopes have not been fulfilled by recent exploration, they have not been completely dashed either!

Watching the Martian life debate is like following an intellectual tennis match; eyes following the ball of alleged evidence as it is served from the court of biology, returned from the court of abiotic chemistry and vice versa. In recent years, this has been especially evident in the debate as to the cause of presence of methane in the Martian atmosphere. Methane is just a trace gas, but in an atmosphere that has achieved chemical equilibrium, it should simply not be present at all. A puff of the gas released into the atmosphere of Mars would be totally destroyed in a matter of years, so its continuing presence there can only mean that something is generating it. Opinion goes back and forth as to the nature of this "something", whether it is biological or some physical or chemical process. Nothing to date definitively clinches the issue one way or the other.

Indeed, recent doubts have surfaced as to whether the Martian methane really exists at all! It may seem hard to credit that the observations pointing so strongly to its presence should all have been misidentifications, but the history of Martian research holds too many similar precedents to say that this is impossible. At the time of writing however, it is simply too early to predict whether these doubts will be upheld or fall by the wayside.

Ever since the *Viking* landers back in 1976 failed to give positive indication of Martian life, the popular perception has been that the result of these experiments was completely negative.

However, the fact of the matter is that one of the experiments—the *Labeled Release* (LR) experiment—actually gave ambivalent results that may even be interpreted as having been weakly positive. This experiment involved a sample of Martian soil being moistened with a nutrient solution containing the radioactive iso tope carbon-14. The idea was that any micro-organisms present in the soil would feed on this "labeled" nutrient solution and expel waste gases carrying the radioactive trace. The presence of life in the sample should be betrayed by a rise in the level of radioactive gas within the experiment chamber. Theory predicted one series of results if no life existed within the soil sample and a markedly different one if life did exist. The trouble is, the real results lay somewhere in the middle!

In the methodology of science, there is a valuable principle called Ockham's Razor. In brief, this states that the explanation of any phenomenon requiring the least number of complications is likely to be true. As life is more complex than inorganic chemistry, most scientists agree that if a phenomenon (such as the LR results) can be adequately explained by chemistry, then chemistry is more likely than biology to be the correct explanation. The difficulty, however, lies with that little word "adequate". The designers of the LR experiment, Drs. G. Levin and P. Straat, have never been convinced that the proposed chemical explanations of their results have really *been* adequate. Levin in particular has always wondered just whose beard Ockham's Razor should be shaving. Chemistry has not (he argues) produced the smoking gun in all the years since the experiment was performed. Biology, on the other hand, can account for the results as long as the number of cells in the sample is small by comparison with, say, the soil in your back garden.

An interesting twist in the continuing analysis of the LR results came in 2001 when J. Miller detected evidence of a cycle in the emission of the radioactive gases. This cycle reminded him of the terrestrial circadian rhythm, except that it was in harmony with the length of the Martian Sol rather than the terrestrial day! If that could be confirmed, it would provide strong evidence that life was indeed present in the soil sample. When dealing with subtle signals in much noise it is, of course, about as easy to read signals *into* the data as *out* of it, and for this potentially very important discovery to stand the test, careful statistical analysis will be required to eliminate this possibility. The jury is still out, but Miller predicts that if the methane source on Mars is biological (and *real*!), a similar circadian rhythm should be found there as well. Accurate atmospheric analysis by future Martian landers might finally answer this question.

In the meantime, Miller and his colleagues at the University of Southern California's Keck School of Medicine have published another ingenious analysis of the old LR data. The team distilled the data into sets of numbers and analyzed these in terms of complexity. As life is complex, the team examined the results in terms of the level of complexity and found that the results agreed far more closely with similar analysis of biological systems than of purely physical processes. Following this analysis, Miller concluded that he was "99 % sure" that the LR experiment had indeed found life on Mars!

Others were not so sure however, arguing that the method has not been sufficiently tried and tested on terrestrial life to act as a reliable test for life on Mars. Once again, more research will no doubt either strengthen or weaken Miller's case, but the final answer (if there ever *is* one!) is not likely to come until further investigation is carried out on the surface of the Red Planet itself.

In general though, it must be said that continuing investigation of Mars has tended to raise rather than lower hopes that some form of life exists there. At least, that has been the trend after the initial nose-dive in confidence following the first *Mariners*. The slow recovery from this nose-dive has been brought about by the recognition that a good deal of water does exist on Mars (even if it is, nowadays, mostly locked up in the form of ice) and that in the remote past, Mars almost certainly experienced conditions more conducive to life than it experiences today. Like Venus, Mars probably possessed surface water (minerals that can only be formed in water have been discovered on its surface) and an atmospheric pressure closer to that of Earth's back at the time our planet was going through its Hadean stage. As we saw earlier, the three planets were almost certainly experiencing rather similar conditions back then and it is a reasonable supposition that if life first appeared on Earth around this time (for which there is evidence, albeit not absolutely compelling evidence) then something similar might have happened on Mars and Venus. Nevertheless, as we said in our discussion of Venus, it would seem that only on Earth did life really come to flourish and, in certain respects, dominate the scene, changing the planet's atmospheric composition and colonizing essentially every part of the globe. If life ever started on the other two planets it either became extinct because of the changing conditions or took hold of a favorable ecological niche and remains there right down to the present day. Given life's resilience in the face of change, the second alternative seems more likely in the opinion of the present writer. Venusian life, we suggested, may have found this niche in that planet's upper atmosphere. In terms of "biological friendliness", the upper atmosphere of Venus looks much more promising than the upper atmosphere of Mars, so if there are any micro-Martians, they are unlikely to be found there. A far more likely environment is underground, maybe from just below the surface (within reach of *Viking's* LR experiment?) to—well—very deep indeed.

A paper by E. Jones and C. Lineweaver of the Australian National University in Canberra and J. Clarke of the Mars Society of Australia, published in the December 2011 issue of *Astrobiology*, specifically addressed this question of underground habitability. With the aid of a model describing how temperature and pressure changes with increasing depth beneath the Martian surface, this team attempted to find the planetary "habitable zone" or region where temperatures and pressure combine to allow the presence of liquid water. The authors find that because of the lower mass of Mars, temperature rises more slowly with depth than it does on Earth—5 °C/km for Mars as against 25 for Earth. The Australian team found that water could remain liquid, as droplets in rock pores, to a depth of around 194 miles (310 km) on Mars and just 47 miles (75 km) on Earth. Experience with terrestrial extremophiles suggests that living organisms of this description can exist between temperatures of –20 °C and 122 °C, the upper boundary being set by the microorganism *Methanopyrus Kandleri*, discovered living in water at 122° along the tectonic plate boundary beneath the Indian Ocean. Thanks to the pressure at the depth of this organism's environment, water raised to these temperatures remains below its boiling point. Jones, Clarke and Lineweaver estimate that sub-surface water within this range might account for as much as 31 % of the overall volume of Mars' liquid water. This volume, according to these authors, constitutes the region of Mars where life is most likely to exist.

Of course, this reasoning is based upon the limits of *terrestrial* extremophiles. It is possible (one might even say "very likely") that Martian extremophiles could very well exist over an even more extensive range. After all, the general prevailing conditions on the planet are more extreme than those on Earth, so any organisms living there may have adapted to conditions that even the more extreme terrestrial extremophiles could not endure.

Moreover, these creatures need not necessarily be single-celled organisms. As recently as 2011, a University of Ghent team led by G. Borgonie and T. Onstott discovered a type of nematode living quite happily in ancient groundwater and feeding on bacteria as deep as 2.2 miles (3.2 km) beneath the surface of the Earth. This little worm-like critter—just one half of a millimeter long—is the deepest-living multi cellular organism yet to be discovered. Because of its deep and dark environment it has been given the name *Halicephalobus Mephisto*, after the demon Mephistopheles ("the one who loves not the light"), to whom Faust sold his soul (Fig. 1.15).

If Martian life does exist, and if it is confined to this planetary underworld, there could be a thriving ecosystem beneath a sterile surface. But maybe this ecosystem could surface in especially sheltered locations. Caves—which are known to exist on Mars—might be good places to search for life without the need for excavation.

A Peter Pan Planet?

Mars, as we have noted, is a small world. Yet, it formed in a region of the early solar nebula that we would normally expect to have contained enough material to form relatively large rocky planets. We might have expected Mars to have been at least the equal of Venus and Earth. So why did it turn out to be the runt of this planetary litter?

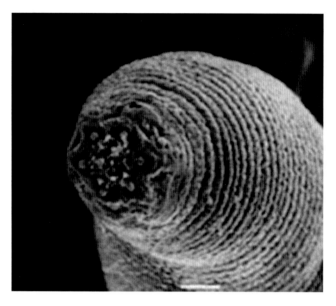

Fig. 1.15 *Halicephalobus Mephisto*. If Martians exist, could they look like this? (Credit: University of Ghent, Belgium)

Planetary scientists nowadays consider Mars to be a planetary embryo; a sort of "Peter Pan" amongst planets that never really grew up. It began to form in the way of its neighboring inner planets, but something stopped it from accreting enough matter to grow up into the equal of Earth and Venus. But what was this "something"? What starved infant Mars of the food necessary for it to grow big and strong like its nearby sisters?

We might suspect giant Jupiter as the culprit. It apparently cleared so much of the material from the gulf of space beyond Mars that no fully fledged world formed there—only a vast host of small asteroidal bodies, as we shall see in Chap. 3. Nevertheless, Jupiter is still a very long way from Mars. We must not be too quick to lay the blame at the feet of old Jove!

A possible explanation has, however, come from an unexpected quarter; from studies into the dynamical evolution of migrating planets based upon simulations conducted at the ALICE High Performance Computing Facility at the University of Leicester in the United Kingdom. These simulations showed that gas giant (Jupiter-like) planets in extra-solar planetary systems are not evenly distributed but tend to favor similar orbits. Further simulations, this time by I. Pascucci at the Lunar and Planetary Laboratory and Richard Alexander at Leicester, revealed that gaps in a pre-planetary disc form at certain distances from the central star due to the evaporation of dust and gas at these locations. The gaps are located in regions too remote from the star for gravity to hold the disc in place against the star's heat, yet close enough for the disc to be too hot to be stable. Within this intermediate region, disc material is hot enough and gravity weak enough to allow matter to escape from the disc in a process known as photo-evaporation. Migrating giant planets are brought to a halt—and may even be sent into reverse—upon reaching these gaps, hence their preference for certain distances from their parent stars. But the importance of these findings for the present subject is that the location of this gap within the early Solar System lies between one and two radii of Earth's orbit from the Sun. Mars orbits right in the middle of the gap, in a region which apparently became quickly depleted of material just as the planet formed. This may well explain the mystery of why Mars never quite grew up!

Incidentally, we might also notice that this depleted zone incorporates much of the region in which liquid water might exist on the surfaces of suitable planets and where advanced forms of life are considered possible. Earth's orbit marks the inner extremity of the zone, so our planet appears to have barely missed being another Mars-like planetary Peter Pan! The absolute distance of both zones will, of course, vary according to the size and radiation output of the central stars, but it does appear that they will vary in unison as the balance of heat and distance from the star is vital to each in its own particular way. Although these are still early days in the study of depleted zones, it may be that their presence reduces the region in which Earthlike inhabited planets can exist. Mars, it is sometimes said, might have been as suitable for complex life as Earth had it been somewhat larger than our planet. Although this suggestion has been disputed (taking into account such complicating factors as the effect on life of the stronger gravity necessary to hold down an atmosphere thick enough to yield Earthlike temperatures at Martian distances) it may be that Mars, and equivalent planets in other solar systems, simply cannot grow sufficiently large for Earthlike conditions to be maintained.

Perhaps they are all relegated to the role of being the Peter Pans of their planetary systems!

Earth's three companions have been thought of as its closest relatives amongst known astronomical objects—the objects most similar in nature to itself. This is quite true, but the fact remains that these nearby worlds also have some striking differences. We have seen some weird things amongst our neighbors! The Sun halting and reversing in its journey across the sky, the same Sun rising in the west and setting in the east during a day far longer than anything experienced on Earth, mountains capped by exotic "snow" at temperatures above the boiling point of water, sulfuric acid hazes, fountains of dust-laden carbon dioxide erupting in spectacular geysers from an icy landscape, landslides that may have been triggered by evaporating carbon dioxide ice and—just maybe—examples of primitive forms of life clinging on in selective ecological niches. But if such relatively Earthlike bodies as these can have their "weird" features, what awaits on those bodies which have little in common with our own dear planet? Let's continue our journey and find out.

2. Giants of Gas and Ice

Immediately beyond the orbit of Mars lies a wide gulf uninhabited by any major planet. It is, however, the abode of a great number of smaller sub-planetary bodies known as "asteroids," about which more shall be said in the following chapter. Further beyond this region, extending to the very edges of the system of major planets, lies the realm of the giant orbs Jupiter, Saturn, Uranus and Neptune. The first two are known as "gas giants" and take us to the limits of the Solar System known to astronomers before the late 1700s. The remaining more modestly proportioned worlds—the "ice giants"—were unknown to the ancients and by comparison with the worlds discussed in the previous chapter of this book are very, *very* far away.

Jupiter

In Greek mythology, Jupiter was the chief of the gods. Not for nothing does the great planet bear this name! It has been said that if an alien were to survey our Solar System, he (she? it?) would probably describe it as a star with one major planet, plus a swarm of minor bodies of little note. Most of the mass of the Sun's system of planets resides in this one massive body and 1,000 orbs the size of Earth could comfortably fit within its globe, but more of this in a moment.

As seen from Earth, Jupiter is regularly the second brightest of the planets. Only Venus consistently surpasses it although very occasionally Mars, at an exceptionally favorable opposition, just manages to outshine it. But Venus is our closest planetary neighbor, so it ought to appear bright! Moreover, it can never be placed opposite the Sun in our skies (the only planet on which that can happen is Mercury) but Jupiter can be visible all night long, the brightest regular nighttime object other than the full Moon that is capable of this feat. No regular star, other than the Sun itself, is as bright as Jupiter as seen in our skies. (I use the word "regular" here to omit the very occasional transitory object which may indeed outshine Jupiter. The past 1,000 years have seen a couple of supernovae become brighter, but they are hardly "regular" denizens of our skies!)

Unlike Venus however, Jupiter does not shine brightly in our skies because it is very close to Earth. On the contrary, Venus at its nearest is over 18 times closer than Jupiter. Jupiter is bright because it is big—really big. We already mentioned how a hypothetical alien might see this planet's place in the Solar System. Let's look at this in more detail. If we had a large enough pair of scales and placed Jupiter in one pan and all the other planets in the other, Jupiter's pan would crash downwards! The mass of this single giant world is some two and one half times greater than all the other solar worlds combined. In terms of size, the average radius of Jupiter clocks in at almost 44,000 miles or nearly 70,000 km, (yielding a diameter of 88,000 miles or 140,000 km approximately). All the other Solar System bodies—beside the Sun itself—could fit within its volume. (The Sun however, could swallow around 1,000 Jupiters, so remembering this fact helps a little to keep things in perspective!).

Through a telescope, Jupiter is a spectacular sight, second only to Saturn amongst the Sun's planets as a source of wonder and amazement. Though it lacks Saturn's spectacular ring system, its disk appears larger (varying between 29.8 and 50.1 seconds of arc in diameter) due in part to its larger intrinsic size but chiefly because of its greater proximity to Earth. Even a small telescope proves sufficient to show the colored belts of cloud across the planet's disk and a scope of moderate size might also be able to pick out one of the true wonders of this world; the Great Red Spot. This is a cloudy atmospheric eddy large enough to engulf at least two Earths and towering some 5 miles (8 km) above the surrounding cloud deck. This great storm has an amazingly long lifetime. It can certainly be identified with a feature appearing in a drawing of Jupiter made in 1859, became unusually prominent in1878 and is widely held to be identical with a feature noted by G. D. Cassini and independently by R. Hooke in the mid seventeenth century. Over the years it has waxed and waned in prominence and to a smaller degree in size, and has varied in color from a rather dull grey to conspicuous brick red. A watch on the Spot reveals something else about Jupiter. It is revolving on its axis—and fast! Watching the Spot progress across the visible disk shows that the planet spins on its axis at a remarkable 28,313 miles (45,300 km) per hour, enabling it to make one full turn of its huge bulk in just under 10 h! This rapid rate of rotation shows itself in the relatively conspicuous flattening of the planet's poles. Adding to the telescopic spectacle of Jupiter are the four bright "Galilean" moons—part of the giant planet's vast satellite family—which perform their constant dance around their primary, at some times passing in front of the planet and casting black shadows on the cloud tops, at others passing behind it or disappearing in the giant planet's shadow or even eclipsing one another. These various phenomena are fascinating (especially seeing one of the moons simply fade out as it enters Jupiter's shadow) and certainly add to the visual spectacle of the giant planet. They are also fascinating worlds in their own right, but any further talk of this must wait for a later chapter.

Once, not too many decades ago, Jupiter was pictured as something like an oversized Earth. That is certainly not to say that conditions there were thought to be very Earthlike—far from it—but the realization that the bodies we call "planets" come in such a wide variety of forms did not really come until the advent of space probes. Prior to that, most astronomers assumed almost as a given fact that all planets had the basic structure of our own. That is to say, a planet was a solid body with a hard and well defined surface. There may or may not be an atmosphere, but if one were present, it would exist as a gaseous envelope clearly defined from the surface, much like the familiar air of Earth. Of course, it might be utterly un-Earthlike in composition or density or temperature, but it would at least mimic Earth's in its basic form. Similarly, if a sufficiently dense atmosphere permitted liquid to exist on the surface of a planet, bodies of this liquid would be basically like the oceans and lakes of home. They may not be made of water, but they would still be recognizable as oceans, seas, lakes or just puddles.

With this in mind, Jupiter was pictured as a solid object, having a well defined surface and presumably topographic features such as mountain ranges and valleys, enveloped in a dense atmosphere filled at high altitude with an abundance of cloud. One model saw it as a rocky world encrusted with layers of ice thousands of miles thick. Other models pictured oceans of liquid ammonia covering much of its surface. The writer recalls one especially vivid description of huge snowflakes of ammonia ice, as large as houses, falling from clouds of the same composition and eventually plunging into what the author described as "that nightmarish ocean" of liquid ammonia. Not a place for a summer holiday to be sure!

Once Jupiter began to be explored by robotic space probes, the picture changed dramatically. From being an oversized and nightmarish Earth, the giant world became something utterly alien. Isaac Asimov was not far wrong when he described Jupiter as a "white hot drop of liquid hydrogen", or words to that effect. Hydrogen is, indeed, the planet's main constituent, comprising around three quarters of the mass of its atmosphere. Helium accounts for a further 24 % with the rest being made up of a variety of substances such as methane, ammonia, hydrogen deuteride, water, ethane, neon and traces of various hydrocarbons such as benzene. Carbon is also present in small amounts, as are ethane, hydrogen sulfide, oxygen, phosphine, sulfur and some silicon-based compounds. The clouds contain a number ices such as frozen ammonia, ammonium hydrosulfide and water.

Imagine descending downward through Jupiter's atmosphere. The first milestone that we pass through is the cloud layer. These are the belts of clouds which give the planet its characteristic appearance in our telescopes and consist chiefly of crystals of ammonia and, maybe, ammonium hydrosulfide. Other materials must be mixed in with these substances as well, as ammonia or ammonium hydrosulfide crystals alone could not produce the colors seen in the cloud belts. It is thought that compounds known as *chromophores* well up from deeper down in the atmosphere and become colorful upon exposure to ultraviolet radiation from the distant Sun. The composition of these chromophores is unknown at present, but phosphorus, sulfur and, possibly, certain hydrocarbons are suspected. The cloud deck is actually very narrow considering the size of the planet and its atmosphere. It appears to be just 30 miles (about 50 km), or thereabouts, thick and to consist of at least two separate layers. The lower, thicker and warmer one is where the colorful compounds occur. Convections cells rise above this layer and, at their summits, clouds of ammonia crystals waft out into formations not unlike the anvil cirrus of terrestrial thunderstorms, obscuring the colorful underlying layers from our view. Swept along by winds of around 225 miles (360 km) per hour, these spreading "anvil cirrus" form the light-colored *zones* so familiar to observers of the planet. Clear regions between these zones expose the deeper, darker and more colorful clouds known to observers as *belts*. There also seems to be a thin layer of water-ice clouds underneath the ammonia layer and it is here where the famous Jovian lightning plays. Thunderstorms discharging lightning bolts a 1,000 times more powerful than those of Earth have been observed within these water clouds.

As we descend lower and lower into the atmosphere, the air becomes ever more dense and the temperature continues to rise. When we reach a level about 620 miles (1,000 km) below the cloud deck, the mainly-hydrogen atmosphere reaches a so-called supercritical fluid state, i.e. a state in which the distinction between gaseous and liquid phases disappears. For convenience, the "atmosphere" can be considered as liquid below this level, but it must not be thought of as an "ocean" or anything as familiar to we Earthlings as that. There is no actual boundary between the "gas" and the "liquid". One simply passes into the other as the hydrogen grows hotter and denser with increasing depth. Rather arbitrarily, the base of the atmosphere is usually considered to be the level at which pressure reaches ten times that of Earth's sea level; a value reached at a depth of around 3,000 miles (5,000 km) from the outermost fringes of the atmosphere. The density at which the supercritical fluid state is achieved is somewhat higher—about 12 times that of the Earth's atmosphere at sea level.

Descending still further, pressure and temperature continue to climb until, at a level almost 10,000 miles (16,000 km) below the clouds, another transition takes place. At a temperature hotter than the surface of the Sun and a pressure of around 200 GPa, the hydrogen starts to behave as if it were a metal. Through this layer, droplets of helium and neon precipitate downward like a weird kind of rain, effectively reducing their abundance higher in the Jovian atmosphere. Then, right at the center of this unbelievable object, there may (or may not!) be a rocky core of between 12 and

45 Earth masses. Not every planetary scientist agrees that a core formed, while others suggest that even if one did form initially, it may have been wafted away via hot currents of metallic hydrogen and diffused through higher levels of the planet's interior. Hopefully, this question will be answered in the not-too-distant future.

A Ringed Planet

Prior to 1978, Saturn was thought unique (at least in the Solar System) in possessing its spectacular array of rings. But that year, distant Uranus was also found to be encircled by a faint ring system and so it came as no great surprise when, in 1979, images beamed back to Earth from the *Voyager I* space probe disclosed a similar faint system surrounding Jupiter. All four giant planets are today known to possess rings systems, suggesting that these features are probably common adornments of giant worlds throughout the Universe. Saturn's rings are unusual only by being so large and bright. By contrast, the rings of Jupiter are very dim except when imaged by space probes looking back at the planet from a point further from the Sun. If the angle separating the Sun and Jupiter is small, forward scattering of sunlight by the ring particles lights up the ring system, rendering it far more conspicuous than it appears from other perspectives. A homely analogy is the way in which a piece of thistle down drifting high in the air becomes visible as a bright point of light as it passes almost in front of the Sun. From Earth's vantage point of course, Jupiter can never be seen between the observer and the Sun, so its rings can never be enhanced through the forward scattering effect for observers on Earth (just as well actually, as that would mean that Jupiter had migrated to some point sunward of Earth and our world would probably have either been flung out of the Solar System or into the Sun. Missing a good view of the Jovian rings is a small price to pay for avoiding either of these fates!) (Fig. 2.1).

Unlike the rings of Saturn, which are comprised of particles of ice, Jupiter's appear to be mostly made up of dust particles. There are at least three components of the ring system; a main ring which, when compared with the other components, is *relatively* bright, an inner circle known as the halo and an outer gossamer ring. The main ring is thought to be made of dusty material ejected from the

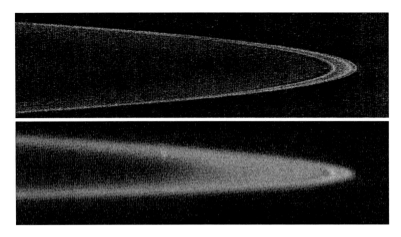

FIG. 2.1 Jupiter's main ring as imaged by *New Horizons* spacecraft (Credit: NASA)

inner moons Adrastea and Metis. Thanks to the powerful gravity of the giant planet, material from these satellites is pulled toward, and thence into orbit around, Jupiter instead of falling back onto the moons themselves. Similarly, material from the moons Thebe and Amalthea probably accounts for the gossamer ring and there is even evidence of a ring of larger, rocky, bodies strung out along the orbit of Amalthea, probably representing debris thrown into space by meteorite impacts on this moon. (Fig. 2.2).

Both the *Galileo* and *New Horizons* spacecraft found something unusual and unexpected about Jupiter's main ring. Within the ring were two sets of spiraling vertical corrugations, one somewhat more prominent than the other. It seems that these were caused by relatively massive objects perturbing the steady state of the ring; the larger of the two resulting from a tilting of the ring out of the equatorial plane by about one and one quarter miles or 2 km. As the spiral patterns were observed to be decaying, the perturbing events must have been very recent. Calculations suggest a date for the formation of the smaller disturbance as sometime during the first 6 months of 1990. The second dated back to 1994. This, we might remember, was the year that a spectacular event—or, rather, series of events—occurred on Jupiter in the form of multiple collisions by a compact group of comets; fragments of the periodic comet Shoemaker-Levy 9 which was tidally disrupted

Fig. 2.2 Diagram of Jupiter's ring system (Credit: NASA)

during an extremely close pass of the planet several years earlier. The spiral pattern within the main ring apparently resulted from impact by material from the broken comet. Presumably, the smaller and slightly older spiral pattern resulted from the impact of debris from an earlier comet that was disrupted close to the giant planet. No observations of this hypothesized object are known and there appears to be no evidence of any sizable fragments impacting Jupiter that year, so this was probably a smaller and less active comet that Shoemaker-Levy, albeit one large enough to leave its fading mark on Jupiter's ring.

Jupiter also boasts a "ring" of a different kind; a torus of gas along the orbit of the volcanic moon Io. As we shall see in a Chap. 4, this moon is the most volcanically active place in the Solar System and it is continually belching out material into the space around Jupiter. But Jupiter also possesses a powerful magnetic field; 14 times stronger than that of Earth. Except for transitory sunspots,

Jupiter is the strongest astronomical magnet in the Solar System. Thanks to this powerful planetary magnetosphere, the gases originating from Io are ionized. Sulfur dioxide is broken down into sulfur and oxygen ions and, mixing with hydrogen ions from Jupiter's own outer atmosphere, form a plasma sheet oriented with the equatorial plane of the planet. Other substances as well boil out from Io and into the plasma sheet, although the origin of all of them is not entirely clear. The yellow-orange glow of sodium was recognized quite early and potassium and chlorine have also been detected. The presence of both chlorine and sodium apparently betrays the presence of sodium chloride (table salt) on Io.

The Jovian plasma sheet is responsible for the intense bursts of radio waves once thought to be due to tremendously powerful flashes of lightning on Jupiter. There *is* lightning on the planet, and it *is* powerful, but it is not responsible for these strong radio emissions. The real reason for these bursts is even more interesting. As Io crosses Jupiter's lines of magnetic field, an electric current as high as two *trillion* watts flows down the lines and into the polar regions of Jupiter. Radio waves are generated through a cyclotron maser mechanism and beamed outward in cones of intense radio emission. In a sense, Jupiter acts as a very weak radio pulsar! When the Earth passes through one of these cones, the blast of radio waves received makes Jupiter appear as a more intense beacon than the Sun itself.

Possessing such a powerful magnetic field, it comes as no surprise to learn that Jupiter also displays spectacular aurora, intensified even further in regions where the magnetic flux tubes connecting Io, and to a lesser degree the other large moons, impinge upon its atmosphere.

About 75 times the distance of Jupiter's radius (approximately 3,000 miles or 5,000 km, in round figures) from the planet, the Jovian magnetosphere interacts with the solar wind, giving rise to a bow shock and elongating "down wind" almost as far as the orbit of Saturn. This makes the length of the "magnetotail" more than 80 % as great as the Jupiter/Sun distance—albeit in the opposite direction! The magnetic field is tilted about 10° with respect to Jupiter's axis of rotation.

The Amazing Shrinking Planet

Jupiter gives out more heat than it receives from the Sun. In effect, it "shines" very faintly in infrared light. Given this fact plus the further knowledge that the overwhelming portion of its mass is in the form of hydrogen and helium, some science writers have described it as being "almost a star" or even "a failed star". True, its composition more closely resembles stars like the Sun than planets like Earth or Mars (its composition is not *exactly* that of the Sun—there is less neon, not quite as much helium and somewhat more of the heavier inert gases—but the overall composition is still pretty close to the Sun's). Yet, its accretion from a protoplanetary disk is not the way stars are thought to form and the fact remains that it does *not* shine in our skies as a second Sun. One astronomer quipped that calling Jupiter "almost a star" is like calling him "almost a millionaire". He is "almost" there in the sense that he earns enough money to live but (being a scientist and not a sports champion or film star) his income still falls far short of the millionaire status!

It is calculated that the material at the core of Jupiter is around six times hotter than the photosphere of the Sun and experiences a pressure of somewhere between 3,000 and 4,500 GPa. This heat is generated, not by thermonuclear fusion as in a star, but by the steady contraction of the planet. Yes, Jupiter is shrinking; at a rate of approximately 2 cm (not quite an inch) each year. It's loss of girth is not going to be noticed anytime soon, but looking backward in time indicates that in its youth the planet had a diameter about twice that of today.

Curiously, if more mass were to be added to Jupiter, it would not become any larger in diameter. On the contrary, there would be little change in physical size until it increased to about 1.6 times its actual mass, after which its present size would be *less*, not greater! To the profound jealousy of dieting humans, the more Jupiter "eats" the "thinner" it becomes! This seeming paradox is the result of the increasing density and compression of matter in the planet's core and it is thought that the present diameter of Jupiter is about as large as a planet of its composition, age and general evolutionary history can become. Younger gas giants can be larger (we have already said that Jupiter itself had a more obese youth) and planets which could be called "bloated Jupiters" have been discovered in orbit around other stars—in orbits very close to the star itself where heating is extreme—but for a planet having Jupiter's age and position, it is about as wide as it can be and it can only slim down further with the passing years.

Adding mass to the planet and thereby further compressing its interior will, however, raise its internal temperature and it is here where the "almost a star" exaggeration is born. If a dozen extra planets of Jupiter's mass were crammed into its volume—so that its mass was increased to 13 times its actual value—temperature and pressure conditions deep within its core would become great enough for fusion of deuterium (the "heavy" isotope of hydrogen) to take place and Jupiter would become a brown dwarf; one of those transition objects between planets and full-blown stars. These could almost be called "missing links" between planets and stars except that, happily, they are no longer missing. They are objects that have thermonuclear fusion, albeit of a comparatively feeble form, at their core in the manner of a star, and dust particles and gaseous methane in their atmospheres in the manner of gas giant planets.

If Jupiter's mass further increased by around 65 times, lithium would also start to fuse in its core. It would still be a brown dwarf, just a heavier and somewhat more active variety. But imagine that Jupiter continues swallowing clones of itself until its total mass reached 75 times its present value. At that point, conditions at the core become extreme enough for hydrogen to fuse into helium and a true star is born. Only then would Jupiter become a genuine star; one of those small, faint, relatively cool but extremely long-lived and very common luminaries known as red dwarfs. So if "nearly" means one seventy-fifth of the mass of the smallest genuine stars, then the great planet could indeed be said to be "nearly a star". In truth though, a factor of 75 is not all that close. Jupiter, for all its "stellar" similarities, remains a true planet, albeit one that from Earth's point of view is really weird in its general characteristics. It is a world without a surface, where the atmosphere and the body of the planet merge without clear distinction, where as much heat comes from the interior depths as from the distant Sun and where giant cyclonic storms that could swallow the Earth and still have room to spare rage unabated for hundreds or maybe even thousands of years. Atmospheric circulation and upwelling convective clouds are therefore driven more by the characteristics of the planet itself (internal heat, rapid rotation) than by solar heating and we can imagine that even if Jupiter were flung into interstellar space, the tremendous thunderstorms and cyclones that churn through its atmosphere would continue unchecked—unlike the Earth where all air would freeze onto a barren surface.

We shall visit the Jupiter system again in Chap. 4 when we look at its extensive family of moons, especially the "big four" Galilean moons visible from Earth in even the smallest of backyard telescopes. But for the present, let's move on to the next of the Sun's gas giants, the magnificent ringed world, Saturn and see what weirdness might also await us there.

Saturn

The sixth planet out from the Sun, the spectacular Saturn, is in many respects a scaled-down version of Jupiter. It is, however, considerably less massive, weighing in at a little over 95 Earth masses. This is quite lightweight compared with Jupiter which, as we saw, comes in just shy of 318 Earth masses. Yet, the discrepancy in the volume of the two giants is a good deal less pronounced. Jupiter could swallow over a 1,000 Earths (1,321.3 to be precise) but Saturn could contain almost 764. This conspicuous difference in mass and considerably less divergence in volume means that Saturn is a planet of very low density. Indeed, of all the worlds in the Solar System, it is the least dense and—at a shade under 0.69 g/cm^3, is the only one having a mean density less than that of water. Hence the oft-quoted statement that, were there an ocean large enough to accommodate it, Saturn would float. (Nevertheless, we will see in the final chapter that it no longer holds the record for the least dense known planet, although we need to leave the Solar System to find those that beat it). The equatorial diameter of Saturn is some 75,335 miles (120,536 km) or nearly 9.5 times that of Earth.

Like the Solar System's largest member, Saturn is primarily composed of hydrogen (about 96 % of its atmosphere) with most of the rest being helium (approximately 3 %) plus traces of methane, ammonia, hydrogen deuteride and very tiny traces of ethane, acetylene, propane and phosphine. As is true of Jupiter, this mainly-hydrogen atmosphere becomes denser with depth, transitioning through a liquid phase of helium-saturated molecular hydrogen and eventually becoming metallic. There are no atmosphere/ocean/solid surface boundaries; no hard and fast dividing line between the various phases. Evidence has been found for a small rocky core (estimated to be between 9 and 22 times the mass of Earth—"small" in comparison to the size of the entire planet), but as is the case with Jupiter, this is not certain. Whatever the truth about a core, the temperature at the center of the planet is very high, albeit not so extreme as Jupiter. Temperatures are estimated to reach a level almost twice that of the Sun's photosphere and (again very reminiscent of Jupiter) heat propagates through the planet such that Saturn actually radiates away into space 2.5 times the energy that it receives from the Sun. The source of this amount of heat is really something of a mystery. Jupiter, we recall, generates its internal heat through a slow and steady shrinkage leading to increasing compression deep within. Saturn is also slowly shrinking and compressing, but that process does not appear capable on its own of supplying the degree of heating that has been observed. It must be getting help from somewhere; some other process that is presumably not active on Jupiter or else is of very minor importance there. An interesting suggestion involves the generation of heat through the friction of helium "rain drops" passing through the deep hydrogen atmosphere on their way toward the planet's center. We saw that helium "rain" is also thought to occur deep within Jupiter, but there is reason to think that Saturn has "rained out" more of its atmospheric helium than its bigger brother. The outer layers of its atmosphere appear seriously depleted in this gas by comparison with the solar abundance and even with the content of Jupiter's outer atmosphere. A continuous helium "rain" could explain both this atmospheric depletion of the gas and the unusually high degree of heat being generated deep down in the bowels of the planet. As to the destination of all this helium—where it finally ends up—a suggestion has been made that it accumulates into a helium shell surrounding the planet's core—if, indeed, such a core exists.

When viewed through our telescopes, the disk of Saturn appears rather bland by comparison with that of Jupiter. Latitudinal markings are faint and except for the rare appearance of a "Great White Spot" (more about these a little later) spots and other definite markings are pale and infrequent. Nothing resembling the persistent Great Red Spot of Jupiter is found on Saturn. It is easy to assume that this blander profile compared with Jupiter implies a quieter and more sedate world and, indeed, that is just how astronomers of an earlier generation interpreted it. But such an interpretation turns out to be quite wrong. As has happened so often since the advent of space probes, in situ imaging has overturned many of our earlier ideas about the other members of the Solar System acquired through years of ground based telescopic observation. Saturn is anything but a quiet and peaceful planet. Quite the opposite. Winds reaching a staggering 1,116 miles (1,800 km) per hour howl through its clouds and thunderstorms unlike anything seen on Earth or even on Jupiter flash and crash within the atmosphere. The calm appearance in our telescopes is a false impression caused by a high haze of ammonia ice crystals that largely block from view the turmoil going on beneath. Lower cloud decks appear to be mostly comprised of ammonium hydrosulfide ice, water ice and at lower altitudes, droplets of liquid water in which ammonia is dissolved. Hydrocarbon chemical reactions also take place in the higher atmosphere triggered by the action of solar ultraviolet light on the small amounts of atmospheric methane present. The products of these reactions are carried downward into the deeper atmosphere in a kind of photochemical cycle that is apparently modified by the planet's seasonal cycle.

Allusion has already been made to the occurrence of gigantic thunderstorms on Saturn. Strangely, these seem to be confined to a narrow "storm ally" at 35° southern latitude. Whether this is a fixed feature on the planet or whether the latitude varies with the season is not known at present. Having a year equivalent to around 30 of ours, the seasons on Saturn are necessarily slow and our close monitoring has not continued long enough to tease out the various seasonal effects that might be present. In any case, the storms might be confined to a particular region of the planet, but they make up for their localized nature with exceptional size, endurance and ferocity. One particular example was followed by the *Cassini* spacecraft from November 2007 until July 2008 and another from January until October of 2009. The diameter of this latter storm was almost 86 % that of the Moon; a colossal 1,850 miles or 3,000 km! The lightning that it sparked was equally impressive, each flash carrying a power equivalent to 10,000 terrestrial lightning bolts. On the other hand, it appears that Saturn sports just one thunderstorm at a time, unlike Earth whose thunderstorms, though tiny by Saturnian standards, number between 6,000 and 8,000 at any given moment. A peculiarity about the thunderheads of Saturn is that they seem to come in two colors. The ones noted thus far have either been very bright or very dark. It seems that Saturn's thunderstorms brook no shades of grey! Clouds of the first variety are thought to acquire their luster from crystals of ammonia ice, maybe in the form of high spreading sheets of cirrus analogous to the anvils that cap terrestrial thunderstorms in their decaying phase. The dark clouds are believed to be rich in carbon, presumably deposited from carbonaceous compounds built up through the action of lightning.

A curious recurring type of feature already touched on in passing is the phenomenon of the Great White Spots. These massive atmospheric disturbances are the most conspicuous features seen on the disk of Saturn through terrestrial telescopes, but they are not permanent features akin to the Great Red Spot of Jupiter. There is no persistent "Spot" on Saturn, but the GWS outbreaks seem to be relatively regular occurrences around the time of summer solstice in Saturn's northern hemisphere. Given the length of the planet's year, that means that they occur about every 30 Earth years and have been noted in 1876, 1933, 1969 and, more recently, in 1990 and 2010. There should have been one in the early 1900s but either it did not happen for some reason or it passed unseen. Maybe it occurred but was not as conspicuous as usual. In any case, a reoccurrence of the phenomenon is again due sometime in the 2030s or 2040s, so (young readers please take note!) a careful watch of the planet then might prove interesting. Considering that Saturn appears quite a deal smaller in our telescopes than Jupiter, these Spots are comparatively more striking than that planet's famous blemish. The one in 1876, for example, was observed with a 2.4-in. (6-cm) telescope and that of 1933 was followed by a very young Patrick Moore using a 3-in. (7.5-cm) telescope. The 1990 example eventually extended into a white band encircling the entire planet.

When the most recent eruption occurred—late in 2010—*Cassini* was in place for a close-up view of the process. From the probe's observation of radio bursts and upwelling of warm gas from deeper atmospheric levels, scientists conclude that this "Spot" (and presumably its earlier counterparts) was triggered by a gigantic thunderstorm deep within the layer of water clouds. This storm burst upward through the visible cloud layers in a way similar to—though on a vastly grander scale—the "overshooting" clouds of powerful supercell storms on Earth; the type which bring devastating weather in the form of large hail and powerful tornadoes. We can hardly imagine what these Saturnain super-counterparts must be like!

Actually, the 2010 storm came some 10 years earlier than had been predicted from the average rate of such events. Following the 1990 "Spot", there had been expectations that another would occur around the year 2020, so the appearance of the most recent one came as something of a surprise. Which only goes to show that predicting storms on other planets is no easier than forecasting them here on Earth!

Even weirder examples of extraterrestrial meteorology are to be found at the poles of Saturn. Infrared images of the planet's south pole reveal a "warm" polar vortex unlike anything else seen in the Solar System. In strong contrast to Earth and Mars, the south pole of Saturn appears to be the warmest place on the planet (excluding the planetary interior of course) with temperatures around 63 °C higher than the planetary average. This is still a shivering -122 °C however, so it is not exactly "warm" by our standards!

Something even less expected exists at the Saturnian north pole; a persisting hexagon! This pattern of cloud extends to about 78° north, appears to rotate at the same velocity as the interior of the planet (almost 10.67 h) and, unlike other clouds, does not shift in longitude. Exactly what causes it is something of a mystery, although most planetary scientists think that it is a standing atmospheric wave pattern of some kind. It has even been suggested that it might be an odd sort of auroral phenomenon, although this seems unlikely.

As expected for a planet basically similar to Jupiter, Saturn also possesses a magnetic field, however its strength in no way approaches that of the king of planets. In fact, it even falls a little short of Earth's at just one twentieth the value of Jupiter's. Nevertheless, the magnetosphere of the planet ensures the occurrence of aurora around Saturn's poles and is known to extend out beyond its giant moon Titan. As we shall see in a later chapter, this moon boasts a very significant atmosphere and ions from the outer fringes of this are wafted off into the magnetosphere, contributing to the plasma trapped there.

The magnetic field is also responsible for sending out radio emissions having a remarkably regular period. These were observed by the *Voyager* flybys in the early 1980s and their period associated with that of the planet's rotation; the rotating orb effectively carries the magnetic field around with it. But as the *Cassini* space probe approached the planet in 2004, new measurements of its period of rotation showed a marked discrepancy from the earlier results. As measured by the radio emissions, the planet's rotation period had increased by more than 5 min since the early 1980s! Further measurements 3 years later confirmed this. Something appeared to be holding back the rotation of the magnetic field. But what could have this effect? How could the magnetic field of a planet be retarded in this way?

The culprit is believed to be another of Saturn's moons—Enceladus. As we shall see in due course, this small moon is best known for its geyser-like ice volcanoes. Water vapor erupted by this object is thought to become electrified and exert such a large drag upon the magnetic field of the planet as to measurably retard its rotational velocity. Strange as it may seem, this tiny moon cheekily tugs at the magnetic coat of its majestic primary!

Planet of Rings

So far, we have mentioned only in passing the one feature for which the planet Saturn is best known; the feature that comes to the mind of astronomer and layperson alike as soon as the planet's name is mentioned. We refer of course, to the magnificent system of rings which gird it (Fig. 2.3).

Long thought unique in the Solar System, the first indication of their existence came as long ago as 1610 when Galileo noted that something looked decidedly odd about the shape of the planet as seen through his tiny refracting telescope. Saturn appeared to have handles—or something that at least looked like handles! Hardly expecting a planet to have handles and innocent of the phenomenon of planetary rings, the pioneer telescopic astronomer opted for the existence of twin moons, one on each side of the planet at very small distances from it. It was not until 1655 that Christian

FIG. 2.3 Saturn and its spectacular ring system (Credit: NASA)

Huygens fully discerned the rings through a larger and more powerful telescope than Galileo's.

Curiously however, there are mysterious hints in the ancient legends of diverse cultures suggesting a prior knowledge of the planet's rings. Many of these (such as the bands around the feet of the statue of the god Saturn at the Roman Capitol) are most probably simple coincidences, but what can we say about an ancient engraved wooden panel from Mexico depicting what appears to be the family of planets, one of which is encircled by a ring (reproduced in Kingsboroughs's *Antiquities of Mexico*, 1830)? And what of the claim that the Maori of New Zealand had knowledge of the ringed planet? E. Best writes in *The Astronomical Knowledge of the Maori, Genuine and Empirical* (1922),

> PAREARAU represents one of the planets. Stowell says that it is Saturn; that Parearau is a descriptive name for the planet, and describes its appearance, surrounded by a ring. The word pare denotes a fillet or headband; arau means "entangled" – or perhaps "Surrounded" in this case, if the natives really can see the pare of Saturn with the naked eye. If so, then the name seems a suitable one.

Best goes on to say that a Maori, commenting on the origin of the name, said that the planet is called Parearau because "her band quite surrounds her." That sounds like the description of an encircling ring but, of course, it is notoriously easy to read all manner of things back into old legends and stories from other cultures. Nevertheless, the apparently widespread allusions to Saturn being "bound" or "surrounded" is curious to say the least.

> **Project 1: The Visibility of Saturn's Rings**
>
> It is well known that Saturn's rings are clearly visible in very small telescopes. A 2.5-in. refractor is enough to give a spectacular view and even large binoculars will show them clearly enough to distinguish their nature. We wonder if, had Galileo somehow known in advance that the planet was surrounded by a ring system, would he have recognized the true nature of what he was seeing in his tiny instrument and stopped thinking in terms of close moons and the like.
>
> How small can an instrument be and still discern the rings? The issue is one of magnification more than aperture as a 2-in. telescope is sufficient with moderate magnifications. But can you detect the rings using 10×50 binoculars (which have the same aperture)? Or 7×50 binoculars? What is the smallest instrument through which you can see the rings?

Leaving aside ancient tales and returning to our present state of knowledge, we know that the diameter of the main ring system has a width of 71,294 miles (114,070 km), from about 4,134 miles (6,630 km) above the planet's equator out to around 75,430 miles or 120,700 km. Yet, despite this considerable width, the rings average only about 65 ft (20 m) in thickness. This makes them, by comparison, thinner than a sheet of tissue paper. They are neither solid nor gaseous, but can be interpreted as a vast system of millions upon millions of tiny micro-moons, ranging in size from around 30 ft (10 m or thereabouts) all the way down to mere specks of dust. The composition of these particles/bodies is principally water ice (accounting for some 93 % of ring material) with the remainder being amorphous carbon and very slight traces of more or less complex organic compounds of the type known as tholins. Beyond the principle or, as we might say, "classical" ring system lies a very faint and distant oddity of a ring tilted at an angle of 27° to the main system and comprised of bodies that orbit

Fig. 2.4 An eclipse of the Sun by Saturn as imaged by *Cassini* spacecraft, 15 September 2006 (Credit: NASA)

Saturn in the opposite direction to those of the nearer and better behaved rings. This black sheep in the Saturnian family appears to be associated with the backward-orbiting (i.e. retrograde) satellite Phoebe; a small and dark body that has also been found guilty of darkening one hemisphere of another of Saturn's moons, Iapetus. Apparently, Phoebe has shed quite a lot of dark dust over the years, not only supplying another ring to the already well endowed planet but also coating the hemisphere of neighboring Iapetus that this moon keeps turned toward it. We will say more about the two-faced Iapetus and the dust-shedding Phoebe in Chap. 4 (Fig. 2.4).

It was earlier thought that the main ring system was primordial; left over from material that failed to be incorporated into the building of the planet billions of years ago. More recent thinking understands the rings to be a lot younger—more like several hundreds of millions of years old—and sees them as consisting of fragments of a disrupted satellite. There is certainly enough material in the ring system to construct a small moon if it all snowballed together. Estimates of the ring system's total mass range from 3×10^{19} to 9×10^{19} kg. One suggestion involves a large moon plunging into the planet and being stripped of its icy shell just before the fatal crash. Be that as it may, some of the ring particles are thought to be of even more recent parentage, namely icy particles ejected by the ice volcanoes of the small but active moon Enceladus, about which more will be said in Chap. 4.

In the smallest of telescopes, the system looks like a single bright ring, but in reality it is rather more complicated than that. It is actually a complex of many rings, separated by divisions and gaps, some of which are visible in quite small telescopes and others revealed only close at hand from space probes. The spaces between rings are not totally void although the particle density there is far lower than within the bright rings themselves. The most conspicuous of the divisions is known as Cassini's division in honor of its discoverer, G. Cassini, who found it in 1675. It is some 3,000 miles (4,800 km) wide and is visible in all but the very smallest telescopes. Indeed, the telescope with which Cassini discovered it had a lens aperture of just 2.5 in. (about 6-cm.) about equal to those incongruously white-tube refractors that became the first telescopes of so many young amateur astronomers of the present writer's generation. The second most prominent is known as Encke's gap and may be detected in telescopes of 8 in. (20-cm) or larger. Despite its official name, it was not discovered by Encke, but by J. F. Keeler in 1888, some 23 years after Encke's death. Encke had contributed so much to the study of the ring system however, that giving his name to this rather prominent ring feature was felt to be right and proper. Lest we are tempted to feel sorry for Keeler, it should be mentioned that his name was also given to a ring gap; one about a tenth the width of Encke's. (As an aside, Encke's Comet—which we shall mention later in this book—was likewise discovered by other people, but was given Encke's name in recognition of his computational work concerning its orbit; work which enabled this object to be re-observed at later returns to the Sun). Encke's gap is located some 83,494 miles (133,590 km) from Saturn's center and is home to the small satellite Pan. This moon appears to have swept the region along its orbit relatively clean of ring material thereby carving out this gap in the rings. The gap is not totally free of material however as the *Cassini* space probe discovered at least three thin knotted ringlets lying within it. On either side of the gap, spiral density waves have been noted. It is thought that these are induced by resonances with other moons orbiting nearby, albeit somewhat further out, than Pan. Pan does, however, induce its own set of spiraling wakes.

Although Cassini's division and Encke's gap are the most prominent divisions in the ring system, space missions—in particular *Voyager*—have identified literally thousands of narrow gaps and thin ringlets within the overall system. Some of these arise from the clearing of ring material by moonlets, such as the clearing of Encke's gap by Pan. Others arise from the sort of resonance between orbiting particles and more distant moons that exist between asteroids of the Main Belt and Jupiter and which create the so-called Kirkwood Gaps within the asteroid system that we shall meet in Chap. 3. Where the orbital periods of ring particles are such that these particles repeatedly come into conjunction at the same places in their orbits, with one of the more distant moons, their orbital eccentricity will be pumped up until they are ejected from these regions altogether. In this way, the regions of resonance are cleared of material. The chief region cleared by this means is the Cassini division. Less severe resonances between ring particles and inner moons create spiral waves through the periodic gravitational perturbations of these moons. Other gentle waves passing through the rings appear to be density waves; small-scale counterparts of the galactic waves that give spiral galaxies their spectacular arms.

In the case of narrow rings, their constituent particles are herded together by the combined action of two "shepherd moons" or satellites located just beyond each edge of the ring. A prime example of this is a very thin outer feature known as the F ring, "shepherded" by the two small moons Prometheus and Pandora (Fig. 2.5).

Speaking of small moons, it should be mentioned that tiny satellites lurk within the rings themselves although only the mere tip of this particular iceberg has been discovered as yet. The A ring, especially, is thought to host moons galore. In 2006, 4 tiny ones were found by the *Cassini* craft. The moons themselves were too small to be imaged directly; what was found were "propeller-shaped" disturbances of the ring particles making distinctive patterns within the ring itself. These patterns were several miles across, but the moons themselves were estimated as being only about 320 ft (100 m) or so in diameter. The following year, a further eight moons were found by the same method and the next year added 150 more. It is estimated that some 3,000 inhabit a zone within the ring about 81,000 miles (130,000 km) out from the center of the planet.

FIG. 2.5 F ring and its shepherd moons, Prometheus (*inside ring*) and Pandora (*outside ring*) as imaged by *Cassini* (Credit: NASA)

There is also evidence of a multitude of moonlets orbiting within the F ring. Odd-looking features that have been given the name of "jets" have been observed within this ring. These come in a wide range of sizes. Large jets stretching for hundreds of miles have been imaged and are believed to arise from small moons colliding with the ring. In 2004, a small moon (designated S2004/S6) was found near the F ring and appears to be pursuing an orbit that periodically makes a passage through the ring. This moon is probably typical of the bodies causing the larger jets. However, in addition to these features, some 570 micro-jets have been noted by the *Cassini* spacecraft up to early 2012. These extend for only a few tens of miles and decay after several hours or, at most, days. Fortunately though, one such micro-jet was serendipitously caught in action while *Cassini* was imaging the shepherd moon Prometheus. This jet extended outward for some 31 miles (50 km) before falling back toward the ring. The distance from the jet's tip to the F ring's core was measured at just under 17 miles or 27 km. Small, briefly existing, jets of this variety are thought to be the results of impacts by very small and icy moonlets (around 0.6 miles or 1 km in diameter) on the F ring. The particular example imaged by *Cassini* is thought to have been caused by one of these moonlets hitting the ring at a very low velocity, indicative of something orbiting close to the ring itself and not coming in from further afield on a more eccentric orbit and higher velocity. According to Nick Attree of Queen Mary University of London, some of these impacting moonlets might be destroyed by their encounters with the rings, presumably bequeathing their icy remains to the F ring in the form of fresh ring particles.

Another surprising finding from data beamed back by the *Cassini* probe is that the rings have an atmosphere! This is not simply an outer fringe of the planetary atmosphere, but a tenuous gaseous envelope surrounding the ring system itself and consisting of molecular oxygen and presumably molecular hydrogen, both derived from the action of ultraviolet radiation on the ring particles which, we recall, are mostly comprised of water ice. It hardly needs comment that this atmosphere is very, very, thin. Indeed, it has been estimated that if was completely condensed out evenly across the ring system, it would form a layer scarcely one atom thick! Although this atmosphere was discovered at close range by a space probe, the *Hubble Space Telescope* has also detected an equally sparse envelope of the hydroxyl radical (OH). Also a dissociation product of water, this does not originate from the ring particles but from molecules of water ejected into space by the ice volcanoes of Enceladus and broken down by the bombardment of energetic ions in Saturn's near environment.

Of Spokes and Spirals

From afar, the rings of Saturn appear calm and stately. Yet, close views from space probes show that changes of various types do occur on timescales of just years. An example of these changes is the evolution of one of three very faint ringlets within the planet's D ring; the dim innermost ring of the system. These three ringlets were discovered in *Voyager 1* images procured in 1980. But when the *Cassini* probe imaged them in 2005, the middle one had not only grown broader but had actually migrated some 125 miles (200 km) inward toward the planet.

Broadening and migrating ringlets were not the only changes having taken place between *Voyager* and *Cassini*. When the first of these spacecraft arrived at Saturn in 1980, a strange and totally unexpected ring feature was noted; one that to this day has eluded a comprehensive explanation. The rings were crossed by radial streaks suitably dubbed *spokes* by the *Voyager* team. These presented a problem in so far as they were remarkably persistent in a way that features resulting from straightforward orbital mechanics should not be. They simply defied an explanation in mechanical terms. A clue to the nature of the material comprising them was given by the way in which they reflected incident sunlight. In *backscattered* light (at small phase angles, more or less opposite the Sun) the spokes stood out *dark* against the bright background of the rings. However, the converse was true at larger phase angles, where Saturn was more of less between the Sun and Voyager. Looking back at the planet, the spokes were imaged as standing out *brightly* against what had become the less intense background of the rings. The changeover from darker-than-the-background to brighter-than-the-background occurred at phase angles of around 60°. At larger phase angles, *forward scattering* of sunlight caused the spokes to significantly brighten, just as the same forward scattering phenomenon causes drifting spider's web and thistledown to light up when it passes more or less in front of the Sun in a clear sky. Now, particles forward scatter light in different ways and to differing degrees depending upon their size, general nature, and composition. Fine sand does it differently to poppy seeds for example. So by analyzing how the material in the spokes looked at differing phase angles, scientists determined that these features were made up of very fine—microscopic in fact—dust particles suspended above the main disk of the ring system. Moreover, the motion of the spokes was found to be essentially in sync with the rotation of Saturn's magnetosphere. These two findings left but one satisfactory conclusion; the particles within the spokes were experiencing electrostatic suspension! But just why that should be and exactly how and why these suspended particles should align into the observed spokes was not clear then ... and it is still not clear today. Some planetary scientists suspect a connection with the incredible thunderstorms that rage in the planet's storm ally, while others think that micrometeoroid impacts on the ring particles might be responsible, but no mechanism is clearly obvious.

As if to complicate matters still further, when the *Cassini* probe began its imaging of Saturn in 2004, not a spoke was to be seen. Then, on 5 September 2005, they reappeared and have become more prominent since then. This suggests that the spokes are a seasonal phenomenon, absent for much of the planet's long year and appearing around the time of the equinoxes. If that is correct, this pattern is obviously telling us something, although precisely *what* remains obscure.

Equally interesting but (we believe!) not quite as mysterious as the spokes are a series of finescale structures found in the D ring. First detected in the gap between the C ring and part of the D, these consist of a series of waves separated by a space of just under 19 miles (about 30 km) but extending a distance of almost 12,000 miles (19,000 km) to the inner edge of the B ring. They appear to be part of a system of vertical corrugations of between about 6.5 and 65 ft (2–20 m) amplitude arranged in a spiral pattern. Interestingly, the period of these waves was found to be decreasing and this allowed scientists to deduce that the pattern itself had its birth sometime in late 1983. We recall that two very similar patterns were found in Jupiter's main ring and one of these appeared to be associated with the Comet Shoemaker-Levy 9 event of 1994. The Saturn ring spiral is thought to have been similarly caused by the impact of a debris cloud having a mass of around 10^{12} kg.; the debris of a small disrupted comet that passed very close to Saturn and was torn to pieces by the ringed planet's tidal effects late in 1983. Like the 1990 Jupiter event, no direct impact evidence is known, but at Saturn's distance any observational evidence of a comet or its impact, even assuming that any sizable portion did actually reach the planet, is not surprising.

Saturn is without doubt a planet of wonder. Here is a giant world that could float on water were there an ocean large enough to accommodate it; a world of incredible winds and enormous thunderstorms, of spectacular rings and myriad moons of all sizes; from the planet scale of giant Titan to the mini-moons tucked inside the propeller-shaped disturbances of the A ring. Some of the moons of Saturn are as strange and mysterious as the planet itself. One has ice volcanoes, another may have its own ring system, one has an atmosphere denser than Earth's, methane rainstorms, seas and rivers of liquid hydrocarbons, and two are seriously considered as possible homes of alien life. We shall return to the Saturn system in Chaps. 4 and 5 and take a closer look at these moons but for now the seventh of the Sun's major planets awaits us. Let's see what weirdness is found lurking there too!

Uranus

With Saturn, we arrive at the limit of the planetary system known to the ancients. Yet, from time to time, a keen-sighted early astronomer may well have spied a small star-like point of light that did not keep to one place amongst the constellations. But if anybody did see it in ancient times, it was apparently not recognized for what it truly was; an extra "wandering star". It was simply too faint and slow moving to betray its true nature.

Coming to more modern times, there is no doubt that observations of the planet were made without its true nature having been recognized. The earliest of which we have knowledge was back in 1690 when John Flamsteed, Britain's first Astronomer Royal, found it but mistook it for a star, actually charting it as 34 Tauri. It seems that Flamsteed observed it at least six times but apparently did not discern its slow motion with respect to nearby field stars. Later, Pierre Lemonnier in France made at least 12 observations between 1750 and 1769. During one sequence of observation, he even saw it on four consecutive nights, yet its motion was too slow for him to recognize. It was not until March 13, 1781 that its true nature began to be suspected. On that night William Herschel found an unknown object that he initially mistook for a comet but which subsequent observations by himself and other astronomers eventually proved to be an additional Solar System planet. Herschel was using a telescope when he found the planet, but it is not difficult to see with the naked eye under clear and dark skies and, on occasions when it traverses a familiar pattern of faint stars, is "discoverable" with the naked eye.

Compared with the long known planets, Uranus is, however, a faint object. But that shyness has nothing to do with its size. Its equatorial diameter of around 31,949 miles (51,118 km) it is just over four times that of Earth. This makes it a pretty big globe, albeit significantly smaller than either Jupiter or Saturn. The real reason for its comparative faintness is its great distance from Sun and Earth. It orbits the Sun at an average distance of around 1,797,924,428 miles (2,876,679,084 km) or about twice that of Saturn. At that distance, the planet takes just over 84.3 Earth years to complete a single trip around the Sun.

In broad terms, Uranus is more akin to Jupiter and Saturn than it is to Earth or Mars and older astronomy books almost inevitably lump the big four—Jupiter, Saturn Uranus and Neptune—together either as "gas giants" or simply as "giants". This is essentially correct, although in more recent times astronomers have been prone to distinguishing between the first two giant planets and these more modest giants, distinguishing the latter as "ice giants". This term should not, however, mislead us into picturing these worlds simply as great balls of frozen volatiles. Like Jupiter and Saturn, Uranus has a deep atmosphere comprised mostly of hydrogen (83 %) and helium (15 %) with around 2.3 % of methane and a trace of hydrogen deuteride. It also contains ice crystals of water, ammonia, ammonium hydrosulfide and methane in its atmosphere, but the "ice" within the body of the planet is, as we shall soon see, not at all like the stuff we find outside on a frosty morning (Fig. 2.6).

The mass of Uranus is approximately 14.5 times that of Earth, making it the least massive of the four giant planets and with a diameter about four times that of our planet, it is the second least dense member of the Sun's planetary retinue, although at 1.27 times the density of water it would, unlike Saturn, sink if dropped into some vast cosmic ocean.

Although there remains some uncertainty as to the precise nature of this planet, what may be called the standard model pictures Uranus as consisting of three layers, namely, a small core of about 0.55 Earth masses and comprised of rocky material, a mantle accounting for 13.4 Earth masses or thereabouts and consisting of a hot and dense fluid composed of water, ammonia and other volatiles grading without definite boundary into a gaseous outer

FIG. 2.6 The bland disk of Uranus (Credit: NASA)

region (an "atmosphere"), principally of hydrogen and helium and having a total mass approximately half that of Earth. The mantle has variously been referred to as an ice mantle or as a water-ammonia ocean, but neither of these terms is used in the same sense as they are in familiar conversation. A hot and dense fluid is not what we think of when we hear the word "ice". By the same token, "ocean" also conjures up equally inappropriate images. On Earth, for example, ocean, land and atmosphere are clearly defined, but no such readily discernible borders exist on Uranus. At some level, the fluid simply reaches a sufficiently low density to be deemed an atmosphere or, from the other direction, the "atmosphere" simply thickens into something more like a liquid than a gas.

This atmosphere is subdivided further into the troposphere where pressures decrease from 300 times that of Earth's atmosphere at sea level to just one tenth of Earth's sea level pressure and temperatures fall from about 320 K in the lower regions to between 49 and 57 K at the highest reaches. The value of 49 K is the lowest recorded not just on Uranus but on any planet in the Solar System. The atmospheric level marked by this temperature defines the tropopause or the boundary at which the troposphere gives way to the stratosphere. The latter exists outward to around 2,500 miles (4,000 km). At that distance, atmospheric pressure is just 0.0000000001 that at sea level on Earth and the stratosphere gives way to the thermosphere or corona; a *very* tenuous mantle of gas, unlike anything associated with other Solar System planets, that extends out as far as 31,250 miles (50,000 km) from the planet's nominal surface. This "surface", by the way, is not a true physical feature of the planet but (in common with the other giants) simply a convenient convention where atmospheric pressure is equal to that of sea level on Earth. It lies within the troposphere, around 187.5 miles (300 km) from that layer's "base" and just over 31 miles (50 km) below the tropopause.

Like its terrestrial counterpart, the troposphere of Uranus is an active region of the planet's atmosphere. Although the disk of the planet does not show the wealth of features displayed by Jupiter or even Saturn, the tropospheric zone is nevertheless marked by complex structures of cloud. It is thought that several cloudy layers exist within this portion of the atmosphere. At levels where the pressure is between 50 and 100 times that of terrestrial sea level, water clouds are believed to occur. Above them, in the 20–40 pressure range, a layer of ammonium hydrosulfide clouds is hypothesized and at still higher levels (in 3–10 pressure range) hydrogen sulfide or ammonia clouds are found. Beyond these again, at levels where pressure has dropped to between one and two times that of Earth's sea level, thin methane clouds have been detected.

For a planet that rotates swiftly (17 h 14 min for the interior of Uranus) we would expect strong winds. We would not be disappointed! Indeed, at latitudes of approximately 60° south, clouds have been seen whipping right around the planet in just 14 h; faster than its rotation. Powerful winds, with velocities as high as 560 miles (900 km) per hour have also been monitored at around 50° latitude in the northern hemisphere. A strange feature of the wind pattern, as determined by tracking tropospheric clouds, is that the more gentle equatorial winds do *not* follow the direction of the planet's rotation. They blow backwards, so to speak! The velocities of these equatorial winds range from 112 to 225 miles

(180–360 km) per hour—not really "gentle" but still a good deal less than those of higher latitudes. The strongest near-equatorial winds blow at latitudes of approximately 20° where, interestingly and perhaps significantly, the lowest tropospheric temperatures have also been recorded. Beyond this latitude, winds change direction and blow with the planet's rotation, increasing in velocity until they peak near 60°. At even higher latitude, velocities fall off again, dropping to essentially calm at the poles themselves.

Conditions within the troposphere (which we might describe as the weather conditions of the planet) appear to have altered since our first really good close-up look through the eyes of *Voyager* in 1986. In some respects, these first detailed views of Uranus were disappointing. Uranus looked very bland; not at all dynamic like the two larger giant planets. Not that it was entirely without detail. A narrow band or "collar" of cloud was observed at latitudes of −45 to −50 and thought to be a denser region of the high layer methane cloud sheet. A bright cloud formation—presumably of the same type—also formed a cap over the southern polar region and some dark bands were evident near the equator. No major "spots" or cyclonic storms were active however, just ten small bright clouds, mostly a few degrees north of the edge of the collar cloud. The planet's northern hemisphere could not be observed during the *Voyager* flyby.

By the beginning of the present century, the far northern region of the planet was beginning to come into view from the Earth's perspective and observations made with the *Hubble Space Telescope* and the Keck reflector initially revealed a very asymmetric planet. The clouds near the south pole were bright, but they had no counterparts near the north. Southern Uranus was bright and northern Uranus dark! But that was about to change. By 2007, following the equinox of the long Uranian year, the southern collar almost vanished while a faint ring of cloud began to take shape around +45°—a northern collar was forming!

Not only that, but in recent years clouds are increasing in number and spots are beginning to appear. Increased numbers of clouds were already being found during the 1990s as new high resolution imaging enabled more careful study of the planet from Earth and as the planet's northern hemisphere came increasingly into view. Most of these early discovered clouds were in the northern hemisphere and it was thought at the time that this imbalance between hemispheres was probably due to the bright collar cloud of the southern hemisphere obscuring any similar features forming there. In consequence, as more of the darker "collarless" northern hemisphere became visible, astronomers had a better chance of finding clouds in that region. Continuing observation proved that this was not the case. The increase in cloud numbers was real, although the clouds in one hemisphere differed from those of the other. Northern clouds appeared sharper and brighter than their southern counterparts and they apparently formed at greater altitudes. Many of the clouds were short-lived (some persisting for just a few hours) although one is suspected of being identical with a feature observed by *Voyager* back in 1986. If that is true, it would appear that some may persist for years, decades or maybe even longer. As mentioned, spots (storms) have started showing up as well in recent years. The first of these—the Uranus Dark Spot as it has been called—was found in 2006 and imaged by the *Hubble Space Telescope*. Two years prior to this, a persistent thunderstorm was noted. Uranus seems to be waking up after presenting its sleepy face to *Voyager* back in 1986. Moreover, activity seems to be moving from the southern to the northern hemisphere as the planet moves away from its solstice of the 1990s. (It should be mentioned that the hemispheres of Uranus experience extended periods of daylight and darkness because of a peculiarity in the planet's axial tilt. But more about that in a little while.)

Beyond the active troposphere, the quieter stratosphere is a region of rising temperatures, from the very cold tropopause to around 850 K at the base of the thermosphere. The heating of the stratospheric layer principally results from the absorption of ultraviolet and infrared light from the Sun by atmospheric hydrocarbon compounds such as methane, ethane and acetylene, although some heat is also conducted downward from the thermosphere. The hydrocarbons themselves are principally confined to a layer between about 62 and 187 miles (100–300 km) altitude with ethane and acetylene tending to condense into a haze layer rather low in the stratosphere. It is likely that this haze is responsible for the somewhat bland face of Uranus, not unlike the erroneous "quiet" impression given to Saturn by that planet's high altitude haze of ammonia ice crystals.

The stratosphere eventually gives way to the thermosphere/corona, a vast but tenuous envelope having a relatively uniform temperature of around 800–850 K. The source of heat for this region remains a mystery. The Sun is simply too far away for its radiation to account for this degree of heating. Neither does auroral activity appear adequate for the task. (Indeed, although auroras do occur on Uranus, activity is much more subdued than on either Jupiter or Saturn). This outermost envelope of the planetary atmosphere consists mainly of hydrogen, but in addition to molecular hydrogen, there are also many free hydrogen atoms that waft off into an extended corona out to the distance of one diameter of the planet from its nominal surface. This feature is, as previously noted, a unique characteristic of Uranus and is a consequence both of the low mass of hydrogen atoms and of the high temperature of the thermosphere. An explanation awaits a full understanding of the reason for this heating. The corona extends so far from the planet that it causes a certain depletion of the ring system (see below) by exerting drag on the small particles orbiting there.

In addition to these atmospheric layers, Uranus also sports an ionosphere extending between 1,250 and 6,250 miles (2,000–10,000 km). Notice that, at these altitudes, the ionosphere overlaps part of the stratosphere as well as the thermosphere. It is actually denser than the corresponding ionospheres of either Saturn or Neptune, possibly due to the concentration of hydrocarbons at stratospheric levels. Its density fluctuates with solar activity, indicating that it is being sustained by the ultraviolet radiation of the distant Sun.

The Reclining Planet!

One of the weirdest things about Uranus is the tilt of its axis of rotation. Unlike the other major planets whose axes of rotation show modest departure from the perpendicular, that of Uranus tilts to almost 98°, so that instead of "standing" in its orbit, this planet "reclines" on its side. Another way of expressing this is to say that the planet's axis of rotation is almost parallel to the plane of the Solar System, but however it is expressed, the situation is certainly very strange and unique amongst the major bodies of the Sun's family (Pluto, as we shall later see, shares a similar reclining attitude, but this world is no longer considered a major member of the Sun's family!). Around the time of the solstice, one of the planet's poles faces the Sun while the other is deep in darkness. At the following solstice, the dark and daylight poles are reversed. Because of the length of the Uranian year (about 84 Earth years) each pole receives around 42 years of daylight followed by 42 years of darkness. What we might think of as a normal day/night sequence is only experienced, at these times, by a narrow strip near the equator and even here the Sun remains very low over the horizon. Around the times of equinox however, the equatorial regions face the Sun and "normal" day/night cycles are experienced over a wider region of the planet. One result of the axial tilt is a greater solar energy input, on average over the period of a Uranian year, at the poles than at the equator. Yet, paradoxically, the equatorial regions are warmer. The reason for this is not known.

The tilt is not just a characteristic of the planet itself. The entire Uranian system of satellites and rings (more about these shortly!) also shares it. Because the satellites and rings orbit the planet at low inclination to its equator their orbits are, consequently, close to perpendicular to the planetary plane. The entire system is seriously askew!

Why does this planet have such an extreme axial tilt? Patrick Moore joked that, according to an old legend, Earth's axis had once been upright but was made to tilt because of the sins of humanity. What must have happened on Uranus? Moore asked! Old tales notwithstanding, a more likely explanation involves a collision between the planet and an Earth-sized body very early in the life of the Solar System. Some astronomers think that this scenario might also explain another Uranian curiosity. Unlike the other giant planets, Uranus radiates very little internal heat (just 1.1 times as much as it receives from the Sun). The reason for this anomaly is not immediately obvious, although some sort of internal barrier to heat deep within has been suggested. Alternatively, the core temperature may have been depleted and most of its primordial heat expelled when the planet was struck by the same giant impactor that literally knocked it on its side.

Curiously, although the rotational axis of Uranus is so strongly tilted, its magnetic field is not. Prior to the *Voyager* encounter in 1986, astronomers assumed that the magnetic field would more or less align with the rotational axis and that the Uranian magnetic poles would also lie very close to the ecliptic plane. That appeared a very reasonable conclusion based on observations of the other planets and the assumption that Uranus behaved as the other planets in this respect.

Reasonable indeed, but totally wrong! Not only is the magnetic field tilted 59° from the planet's rotational axis, but the field is not even centered on the planet's core. The center of the field is displaced from the planetary center by almost one third of the planet's radius in the direction of the south pole! One result of this is a highly lopsided magnetosphere with the magnetic field strength at the "surface" being stronger in the northern than in the southern hemisphere. A further result is the occurrence of aurora far from the high latitude regions. Because of the "sideway's" rotation of Uranus, its extended magnetosphere (magnetotail) is twisted like a corkscrew as it extends away from the planet in the anti-solar direction. Protons and electrons, together with a small number of H_2^+ ions, probably emerging (at least in part) from the extended corona become trapped in the magnetic field and continually pepper the moons and ring particles, probably contributing to the dark color of the latter via a relatively rapid pace of space weathering.

Although the oddly tilted magnetic field seemed initially to be a strange departure from the other planets, this peculiarity was later found to be shared by the other "ice giant", Neptune and may well be a common feature of worlds of this type. It has been suggested that, whereas the field of planets as diverse as Earth and Jupiter originate within the core of these worlds, those of the ice giants might be generated within the water/ammonia "ocean" or at some depth not far below the nominal surface. For the present however, these off-center fields continue to be mysterious.

Another Ringed Planet!

The discoverer of Uranus, William Herschel, maintained that this planet possessed rings similar to those of Saturn. He claimed to have detected the ring in 1789, describing it as red in color and even deduced its angle relative to Earth. Unfortunately, because Herschel's ring system was never confirmed and never acquired wide acceptance amongst astronomers, the rings of Uranus were by and large forgotten (Fig. 2.7).

Fig. 2.7 Uranus, showing atmospheric clouds and rings (Credit: NASA)

This all changed in 1977. On March 10 of that year, Uranus was predicted to occult (eclipse) a star catalogued as SAO 158687 and a team of scientists—James Elliot, Edward Dunham and Douglas Mink—planned to take full advantage of this opportunity by observing the occultation from the Kuiper Airborne Observatory. Needless to say, they were not seeking planetary rings, but occultations of stars provide good opportunities to study a planet's atmosphere as the light of the star briefly shines through the gaseous envelop before being eclipsed by the opaque disk and briefly acts as a probe of the outer regions of the atmospheric mantle. Moreover, because star positions are known with a high degree of accuracy, very accurately timed occultations enable precise positions of the occulting body to be determined. Upon analyzing its results, the team found that the star had unexpectedly vanished for a brief time just before the expected occultation began. One of the scientists quipped that it must have passed behind the ring. A true word spoken in jest as it turned out. Further examination revealed that the star had blinked out five times both before and after the occultation. The first suspicion that a new satellite had been discovered was quickly dismissed. All joking aside, the team really *had* discovered a Uranian ring system! Additional rings were later detected and the system eventually imaged by *Voyager* in 1986. The space craft also discovered an additional two rings, increasing the total to 11.

The ring system surrounds the planet's equator and shares its unusual tilt, as we earlier remarked. The rings are optically faint, in part because the particles comprising them are grey in color, possibly (as mentioned above) due to space weathering by charged particles within the planet's magnetosphere. The particles are also quite large (perhaps "bodies" rather than "particles" would be a better word to describe most of them), the majority ranging in size from about 8 in. to some 65 ft (20 cm to 20 m), with very little fine dust to reflect incident sunlight. The rings are actually quite opaque, a fact aiding their 1977 discovery. The main ring system extends from about 24,000 to 61,000 miles (38,000–98,000 km) from the planet, although each of the separate rings is individually only a few miles wide. This narrowness is a little odd and seems to imply the existence of "shepherd moons" constraining the spread of the debris that forms the rings. However, suitable shepherds have only been found for the fifth ring from the planet, in the forms of the small moons Cordelia and Ophelia. Similar moons are believed to exist for the others as well, but have thus far eluded discovery. If the other shepherds continue to be elusive however, it may be back to the drawing board for an explanation of the small width of the rings (Fig. 2.7).

The ring system is thought to be quite young by the standards of such things, probably only about 600 million years old and to

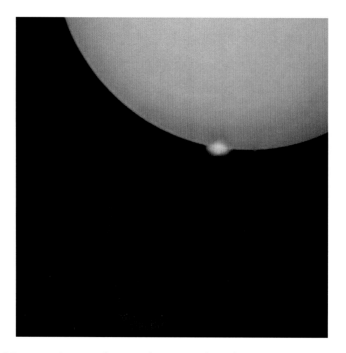

FIG. 2.8 Uranus, rings and auroral activity (Credit: NASA)

represent the remains of two or more moons that once orbited in this region before coming to grief in a destructive collision.

Far from the planet, at over twice the distance of the main ring system, a pair of rings forming an "outer" system was discovered by the *Hubble Space Telescope* in 2005, bringing the total number to 13. The *Hubble Telescope* also found two small moons sharing the orbit of the outermost ring. Unlike the inner rings, these distant ones are not drab grey in color. The innermost is red and the outermost blue, according to observations made in April 2006 at the Keck Observatory. It is thought that the outermost (blue) ring might be composed of minute particles of water ice originating on the surface of one of the satellites (now known as Mab) discovered in 2005. If small enough, ice particles can scatter blue light giving a ring of suitable particles the required color (Fig. 2.8).

Does the discovery of a ring system finally vindicate Herschel's observations? Not really. The brightness, color and even the orientation of his reported rings do not square with the ones found in recent years. Presumably, Herschel was deceived by reflections within the telescope or some such phenomenon and simply read too much into too little. He was not the first to do that, nor was he the last. It is one of the perpetual hazards of visual astronomy. A little like the Martian canals … although in this instance, Herschel was closer to the mark. Uranus at least *does* have rings, even if not the ones he thought he saw!

Neptune

Being accustomed to drawings or models of the Solar System in which the major planets are represented by globes sitting astride neat concentric rings around the yellow disk or globe doubling as the Sun, we are apt to (subconsciously) think of it as a relatively compact family of worlds were neighbors orbit quite close together. But diagrams and models are deceptive at this point. Not deliberately of course, but inevitably. If the Solar System were drawn to scale or a scale model made for display purposes, either the orbits of the inner planets would be too small to see or the display would become too large to be practical. The truth is, the Solar System is a very big place!

The planet Uranus, as we have just seen, orbits the Sun at a distance of close to 1.8 *billion* miles. That is over 19 times the distance from Earth to Sun. But beyond Uranus, there is a gap of just over another billion miles before the next planet, Neptune, is reached. The average distance of this remote world is some 30 times larger than the space between Earth and Sun (Fig. 2.9).

Neptune differs from the other major planets of the Solar System in being the only one that is not visible with the naked eye (although a good pair of opera glasses will show it) as well as being the only one whose discovery was not a matter of accident. At least, in so far as its "official" discovery is concerned because, like Uranus, observations of the planet were made before its true nature was realized. It appears in two of Galileo's drawings made on December 28, 1612 and January 27, 1613. Galileo's principal subject of the drawings was Jupiter, but one of the background "stars" that he depicted has since been identified as Neptune. It just happened to be close to Jupiter in the sky at that time. Oddly, it appears that he was aware of a slight movement in one of these "stars" between his drawings, but apparently did not follow this up any further. Galileo's other astronomical discoveries caused

FIG. 2.9 Neptune, showing signs of atmospheric activity as imaged by Voyager 2, 16–17 August, 1989 (Credit: NASA)

a big enough stir amongst the intellectuals of his day, so maybe finding another planet would have been too much to take in—but it would be interesting to know how the history of astronomy may have developed had Neptune been recognized before Uranus!

Real astronomical history, of course, took another and possibly even more interesting turn. To briefly reiterate the well known story, Alexis Bouvard noticed, following his 1821 publication of the orbit of Uranus, that this planet was not behaving quite as he had predicted. Continuing observations revealed that Uranus was drifting slightly from its predicted path and this led Bouvard to hypothesize that another planetary body must be gravitationally perturbing its orbit. Working from this hypothesis, John Couch Adams made several calculations—between the years 1843 and 1846—of the possible path of the proposed new planet. Working independently, Urbain Le Verrier also made several calculations during the years 1845 and 1846 and when these were published, Adams managed to persuade a seemingly skeptical James Challis, director of Cambridge Observatory, to conduct a search for the planet. This search apparently proved fruitless, but in the meantime—across the English Channel—Le Verrier successfully persuaded Johann Gottfried Galle of the Berlin Observatory to conduct a search of his own. One of Galle's students, Heinrich d'Arrest made the suggestion to Galle that a recently drawn star chart of the requisite region of sky should be examined in the hope of finding a "star" that did not remain at a fixed position. This procedure paid quick dividends, as that very night (September 23, 1846) the planet was found by the Galle/d'Arrest team just 1° from the Le Verrier and 12° from the Adams predicted positions. Back in England, Challis belatedly realized that he had actually observed the planet on August 8 and again on August 12 without realizing it. His lack of an up-to-date chart, and (dare we suggest it?) his apparent lack of enthusiasm for the project cost him the recognition for an important discovery in planetary science.

Just 17 days after the "official" discovery of Neptune, the planet's largest moon (Triton) was found by William Lassell, but the planet is so faint and distant that no obvious detail could be discerned on the disk itself. Prior to the advent of space probes, space-based telescopes and adaptive optics, little could be learned about Neptune and it was generally assumed that it was a slightly smaller and more remote twin of Uranus (and there was not a great deal known about *that* planet either!). That was quite a reasonable assumption and, broadly speaking, it is true enough. Nevertheless, Neptune differs from Uranus in some important respects and, like every planet examined thus far, is very much its own person.

The planet resembles Uranus in so far as both are classified as "ice giants", using "ice" in the rather extended manner that we have already spoken about when Uranus was discussed. In common with Uranus, Neptune has a deep atmosphere estimated to contain between 5 % and 10 % of the planet's total mass and to extend, possibly, as much as 10 % or 20 % of the distance inward toward the presumed planetary core. At high altitudes, this gaseous envelop consists of 80 % hydrogen, 19 % helium and some methane. This last gas absorbs red light and its presence in approximately equal quantities in the atmospheres of both Uranus and Neptune plays an important role in giving these worlds their blue coloration. The color of Neptune is, however, more pronounced than the pale icy blue of Uranus, possibly indicating that something in addition to methane is responsible for its color.

Project 2: Colors of the Ice Giants

Images beamed back from space probes and from the *Hubble Space Telescope* clearly reveal the ice giants to be, like Earth though for very different reasons, "pale blue dots"—indeed, not altogether that "pale" in the case of Neptune! Yet, in many books on observational astronomy, we see these planets being described as appearing *green* in optical telescopes. Patrick Moore speaks of the "dim greenish disc" of Uranus as seen in small telescopes and one handbook of observational astronomy describes Uranus as "a tiny green object" and Neptune as having a "faint greenish color" as seen in modest instruments. Some publications have even referred to these worlds as the "green giants". Why *green*? Why not *blue*? ("Bluish-grey" or "grayish-blue" have been about the closest to Neptune's real color that this writer can recall seeing published in the literature of visual planet observing).

Our atmosphere is not as transparent to the shorter wavelengths of light. This is a good thing in so far as it protects us from dangerous short-wave ultraviolet and harder radiation, but it also depletes the blue light from astronomical sources. The Sun is a good example. To us, observing from the bottom of Earth's ocean of air, the Sun appears yellow. Yet, from outer space, it shines with a bluish-white radiance. Here on the ground, fluorescent lamps marked as 6,000 K shine a light which is clearly a little on the blue side of pure white, yet 6,000 K is also approximately the temperature of the Sun's photosphere. Take a 6,000 K lamp hundreds of miles above the Earth's surface and view it through a telescope, and it would also appear yellow—just like the Sun and other G-type stars. Now, returning to the ice giants, their atmospheres absorb the longer wavelengths (red light) thanks to the presence of methane and whatever else it is in Neptune's atmosphere that adds to the absorption of these wavelengths. The light reflected back from these planets is therefore bluer than incident sunlight, hence their color. However, as this light passes through Earth's atmosphere, blue wavelengths are absorbed

Continued

> **Project 2: (continued)**
> more than light of longer wavelengths and the perceived color of these planets is—in a manner of speaking—reddened; just as the Sun is "reddened" from blue-white to a shade of yellow. But because the light reflected from these planets is bluer than direct sunlight, the "reddening" does not go as far down the spectrum as yellow. It stops one color short—at green.
>
> What color do they appear to you? Do you see much difference in their respective colors? (In comparing their colors, remember that Uranus is about five times brighter than Neptune. We cannot detect color in faint objects, which is why neither planet looks very colorful in, say, a pair of small binoculars. Neptune appears in a 12-in. telescope about as bright as Uranus does in a 6-in., so color comparisons made with the same instrument may be less reliable than when a larger aperture is used for Neptune.) Do the colors alter appreciably when the planets are observed at lower elevations?
>
> As Uranus is a paler blue than Neptune, we may expect it to appear more of a yellowish-green while Neptune is a more definitive green. Is that true or incorrect in your experience?

At deeper levels, this gaseous component merges (as is typical with giant planets, without any well defined boundary) with a mantle of "ices" in the atypical sense of the word used in planetary science; that is to say, with a hot, highly dense fluid of water, methane and ammonia existing at temperatures between 2,000 and 5,000 K. The total mass of this mantle is estimated to be some 10–15 times that of the entire planet Earth. It has been theorized that at a depth of nearly 4,400 miles (about 7,000 km) methane decomposes into carbon and hydrogen. Thanks to the prevailing high pressure and temperature, the carbon component is thought to take the form of crystals of diamond, which then fall toward the core as continuous diamond sleet. Other weird processes are hypothesized to occur within the mantle as well. It has been hypothesized that there may be a layer of ionic water in which the water molecules are broken down into a soup of ionized oxygen and hydrogen atoms. Still deeper in the mantle, there may even be a region of "superionic water" where oxygen crystallizes and hydrogen ions float around freely within this crystal lattice of oxygen!

Also in common with Uranus, Neptune is believed to have an iron-nickel-silicate core thought to be about 1.2 times as massive as the Earth. The pressure at the center of this core—the very center of the planet itself—is estimated to be about seven million times greater than that of Earth's atmosphere at sea level and to possess a temperature possibly as high as 5,400 K.

The magnetic field of Neptune also has close resemblances to that of its nearest neighbor. Just as the Uranian field is tilted with respect to the rotational axis of that planet, so the Neptunian field is likewise tilted—to some 47° relative to its rotational axis and well offset (by some 8,438 miles or 15,500 km) from the planet's center. This discovery, made by the *Voyager 2* probe, came as a surprise to planetary scientists. The strange arrangement of the Uranian field was already known but was assumed to be a consequence of the extreme axial tilt of that planet. The strong similarity between the Uranian and Neptunian magnetospheres was not expected and has caused scientists to rethink their hypotheses as to how the magnetic fields of these two planets are generated. The electrical conductivity of the water-ammonia-methane fluid probably gives a clue and it is thought that convective motions within a thin spherical shell of this fluid might be the source of the ice giants' weird magnetic fields.

Thus far, the two planets do look like identical twins, but on closer inspection a number of differences begin to show.

For one thing, Neptune is a little smaller than Uranus (in rounded figures, 30,000 miles or 50,000 km as against 32,000 miles or 51,000 km for Uranus) but is somewhat more massive. Neptune weighs in at 17 Earth masses as against 15 for Uranus, meaning that Neptune is the denser of the two ice giant planets. A greater difference lies with the axial tilt of the more remote world. Contrasting with the reclining Uranus, Neptune stands relatively straight with a tilt of 28.32°, not greatly more than those of Earth and Mars (23° and 25° respectively). Neptune, therefore, does not share the seasonal craziness of Uranus but goes through seasons not unlike those of Earth and Mars although, of course, on greatly extended timescales owing to a "year" close to 165 times longer than that of Earth.

One common feature shared by both Neptune and Uranus, ironically, betrays a difference and not a similarity between these two planets. This is the temperature at the nominal surface (i.e. where the atmospheric pressure equals Earth's at sea level). The two planets are equally warm, even though Neptune lies half as far again from the Sun as Uranus and receives only 40 % as much solar radiation. Uranus radiates only a trace more heat than it receives. Neptune radiates about 2.61 times as much. Exactly why this is so is a mystery, but it is possible, as suggested earlier, that the peculiarities of Uranus—the low density, relative lack of internal heat and the extreme axial tilt—are related and may stem from the massive impact which is thought to have occurred early in the life of that planet. The axial tilt is the most obvious result of this, and the lack of such an extreme tilt of Neptune probably indicates that it did not experience a similar impact in its youth. Its development may therefore have been more "normal" for an ice giant just as Venus, rather than Earth, may be a more "normal" terrestrial planet thanks to its apparent avoidance of the sort of collision that is widely believed to have given Earth its Moon.

Neptune also appears to be a stormier place than Uranus and is the site of the strongest winds yet recorded within the Solar System. The Neptunian troposphere, or lower atmosphere where temperatures decrease with altitude, extends to levels where the pressure is one tenth that of Earth's atmosphere at sea level. This marks the tropopause and the beginning of the stratosphere, which in turn gives way to the thermosphere and exosphere. Like the other planets, the troposphere is where the meteorological action takes place. Wild storm systems have been tracked packing winds of some 1,230 miles (2,100 km) per hour! It is thought that several distinct layers of cloud having differing compositions are also present within this zone. The upper layer occurs where pressures are below that at Earth's sea level and consist of methane. Lower down in the atmosphere—at pressures between one and five times that at sea level—ammonia and hydrogen sulfide are believed to form further cloud layers. At pressures of five times sea level there may be clouds of ammonium sulfide, hydrogen sulfide, water and ammonia. High altitude clouds have been seen casting shadows on the

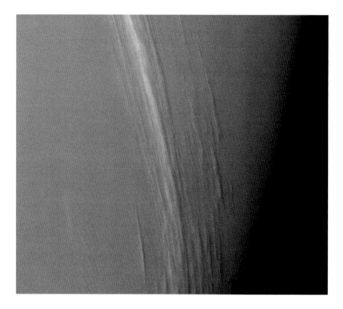

FIG. 2.10 *Clouds* in Neptune's atmosphere (Credit: NASA)

underlying cloud banks and there are bands of the former some 31–94 miles (50–150 km) wide lying about 31–69 miles (50–110 km) above the main cloud deck. These bands circle the planet at constant latitudes. Prevailing (non-storm) winds at the cloud tops vary in velocity from about 870 miles (1,400 km) per hour at the equator to "just" 560 miles (900 km) per hour at the poles. Most of the winds move in the *opposite* direction to Neptune's rotation.

The lower stratosphere is somewhat hazy due to various products, such as ethane and acetylene, arising from the ultraviolet dissociation of methane. Trace amounts of carbon monoxide and hydrogen cyanide have also been found in this atmospheric layer. Despite its much greater distance from the Sun, the stratosphere of Neptune is actually warmer than that of Uranus; a curious finding explained by the higher concentration of hydrocarbons absorbing sunlight and converting it into heat (Fig. 2.10).

Yet Another Ringed Planet!

As soon as the discovery of Neptune was announced, it became the focus of attention for some of the most experienced astronomers of the day. Such was William Lassell, who discovered the planet's largest moon, Triton, on October 10, just 17 days after the official discovery of the planet itself. However, Triton was not all that Lassell suspected that he had found. On October 3, he wrote in his journal "I observed the planet last night and suspected a ring … but could not verify it." Come nightfall on that same day, he "showed the planet to all my family and certainly tonight have the impression of a ring." Once more, on October 10, he observed the ring and wrote that he could even discern the background sky through it. By then, almost all doubt as to the ring's reality had vanished from his mind—*almost* all doubt, not quite all! His later observations failed to confirm these early ones and his confidence in the ring's reality faded. In the meantime though, several other experienced and skilled astronomers—Dawes, Nasmyth and John Hartnup to cite three prominent names—had "confirmed" the existence of something strongly resembling Saturn's rings when viewed from an edge-on perspective. Another well known astronomer, J. Hind, did not at first detect any ring surrounding the planet, but after it was apparently seen by other observers, he seems to have taken a more careful look and likewise detected something resembling an edge-on ring.

Alas, subsequent history was not kind to the Lassell ring. Later observations of Neptune with larger telescopes (including some made by Lassell himself) failed to confirm its existence and the whole thing just faded away into time. A mixture of preconception and minor flaws in the telescopic optics is now thought responsible for the entire episode (Fig. 2.11).

Nevertheless, the non-existence of the Lassell ring did not close the saga of Neptunian rings in general, although it would be many years before the subject was raised again. Ironically, data suggesting a ring was gathered during an occultation of a star of approximately equal brightness as the planet in 1968, but this went unrecognized until after the Uranian rings were discovered 9 years later. If Uranus had rings, then the possibility of a Neptunian ring system was once again an open question and the 1968 occultation data reexamined. Evidence for what appeared to be a partial ring or arc was found within the data, but it was not definitive. Another occultation took place in 1981 and a Villarova University team led by H. J. Reitsema noted a dip in the star's light while still separated from the planet's disk, suggesting the presence of a ring

112   Weird Worlds

Fig. 2.11 The rings of Neptune (Credit: NASA)

or at the very least a ring arc. This proved not to be so however. What this team had witnessed was a highly improbable event; a secondary occultation by a small and at the time unknown Neptunian moon! This body was later observed by *Voyager* and has since been named Larissa. Careful observation of another occultation on September 12, 1983 suggested a possible ring, but later occultations yielded conflicting results. Some suggested the presence of a ring, others did not. Was this evidence of a broken ring or a system of disconnected ring arcs?

The issue was finally settled when *Voyager* arrived at the planet in 1989 and imaged a definite ring. Subsequently, the ring has also been imaged by the *Hubble Space Telescope*. The rings, as *Voyager* imaged them, are quite faint and have a clumpy structure, possibly indicative of small moons orbiting close to or even within them. The outer ring—the one responsible for the inconsistent occultation results—contains five individual arcs. This is known as the Adams ring, named in honor of the English scientist who mathematically deduced the presence of the planet from perturbations of the orbit of Uranus. It is somewhat ironic then, in view of the rivalry between this Englishman and his French counterpart, that the five arcs comprising this ring should be named Courage,

Giants of Gas and Ice 113

FIG. 2.12 Arcs visible in Neptune's rings (Credit: NASA)

Liberte, Egalite 1, Egalite 2 and Fraternite (Courage, Liberty, Equality and Fraternity)! The arcs initially seemed difficult to explain as one would expect their composite particles to have long ago spread out around their collective orbit and the arcs smeared together into one continuous ring. It now appears that they are kept confined by the gravitational influence of a small shepherd moon orbiting a little distance inward from the ring itself. This moon, known as Galatea, is also responsible for 42 radial "wiggles"—each having an amplitude of about 19 miles (30 km) within the ring. Nevertheless, although Galatea keeps the arcs from quickly disappearing, it can probably not hold back the inevitable for very long. Changes in the ring detected by ground-based observations from W. M. Keck Observatory in 2002 and 2003 suggest that one of the arcs (Liberte) may be gone in as little as 100 years. This further implies that the present form taken by the Adams ring has not existed for long and that the system appears to be very dynamic and in a constant state of flux on timescales as short as centuries (Fig. 2.12).

Contrasting with the Uranian rings, those of Neptune contain a good deal of micron-sized dust particles. In this, they more closely resemble the ring system of Jupiter. The particles appear to be dark and probably consist of a mixture of ice and organic compounds, probably debris from a small disrupted inner satellite of the giant planet. They vary in width from about 60 miles (roughly 100 km) or even less to nearly 3,000 miles (5,000 km). The innermost ring (known as the Galle ring in honor of the planet's discoverer) circles Neptune from 25,600 miles (41,000 km) to nearly 26,900 miles (about 43,000 km) above the planet itself. From these dimensions, it can be seen that this is one of the wider rings in the system. By contrast, the outermost Adams ring has a width of just 9–31 miles (about 15–50 km) and is located almost 40,000 miles (63,930 km) from the planet.

Where Do the Ice Giants Fit in to the Solar System?

One of the weirdest things about the two ice giants is that they are there at all! In the days of astronomical innocence, that is to say prior to the discovery of other solar systems in the 1990s, it was generally assumed that the formation of the Solar System was an orderly and regular event (to suggest anything else smacked of the heresy of catastrophism!) and the planets formed pretty much in the regions where they are located today. On a cursory examination of the System, this did not seem too bad an idea. The portions of the cloud of gas and dust that would become the system of planets should have been thinner closest to the infant Sun thanks to the depletion of dust and volatile materials caused by a combination of solar radiation pressure and higher temperatures in that region. At what we might call "middle distances" the pre-planetary nebula would have been at its densest and cool enough to be rich in volatiles. At large distances, the nebula would have thinned out again, eventually blending into the extremely rarefied matter of interstellar space. Broadly speaking, this seemed to account for the general properties of the Solar System; small and rocky planets closest to the Sun, Gas giants at middle distance decreasing to lesser giants (or ice giants as we now call them) at large distances, dwindling off to diminutive Pluto on the fringes (assuming that this was not an escaped Neptunian moon; a rather widely held hypothesis last century).

Closer and more detailed examination, however, gave a less satisfactory result. Of particular relevance for us, the ice giants seemed just *too* far away to be readily explained. Assuming (in line with the prevailing orthodoxy of uniformism) that no great disturbing events had occurred to eject these worlds to their remote habitations and that, in consequence, where we see them is where they formed, theorists came up against great difficulties in explaining how they could have grown to be as large as they are. Certainly, they are significantly smaller than Jupiter and Saturn, but it was difficult to see how there could have been sufficient material at their large distances from the Sun to permit the formation of *any* sizable planetary body.

Then, in the final couple of decades of the twentieth century, accepted orthodoxy on these matters fell apart. Pluto, it turned out, was not a fully-fledged planet at all, merely one of a myriad of small bodies, some of which may actually outweigh it. More devastating though, the discovery of other planetary systems brought with it the shocking revelation that our Solar System was not the archetypal model of all planetary systems but just one of countless forms that these systems can assume. And some of these (indeed *most* of these!) looked pretty weird from our perspective. So weird in fact that they failed to fit the regular, uniformist, model that had been widely accepted for decades. The feature of these alien planetary systems that was hardest to fit into the popular model was the existence of so-called "hot Jupiters" or massive gas giant planets orbiting their primary stars at very small distances. There just seemed no way that something as rich in volatiles as Jupiter or its extrasolar equivalent could form closer to its star than Mercury is to our Sun. Yet hot Jupiter after hot Jupiter was being discovered; some so close to their central star as to make the orbit of Mercury look distant by comparison. Not content with that, discoveries of other systems very quickly brought to light another class of planets almost as shocking as the hot Jupiters. These were the "eccentric Jupiters"; gas giants moving through their systems on orbits more reminiscent of those followed by periodic comets within the Sun's family. Clearly, our thinking about the genesis and evolution of planetary systems had to undergo a major readjustment!

This readjustment involved a paradigm change from belief in an orderly evolutionary process to one in which violent, catastrophic, stochastic events played a critical role. Planets did not simply form at one place in a pre-planetary nebula and stay put like obedient children. The process was also far from neat in the sense that there was much material left over after all the planets within a system had formed. Moreover, this material was not to be dismissed as of no importance, nor did it consist entirely of small and insignificant bodies. Much "left-over" material existed in the form of bodies the size of small planets (Pluto-like objects) and even the smaller bodies that numerically made up the lion's share of this material existed in such enormous numbers as to play an important role in the way the entire system turned out in the end. In short, the picture that emerged was one of planetary systems in which planets gravitationally interacted with each other and with the myriad small objects surrounding them to such a degree that some were hurled into interstellar space, some sent crashing into the central star and others caused to migrate inward until they reached distances from their star far smaller than those of any planet in our own Solar System. Yet others experienced such close encounters that the eccentricity of their orbits increased to a degree exhibited only by comets within the Sun's family. There even appeared to be evidence that some of these eccentric planets were the outcomes of planetary collisions and mergers between gas giants.

We will see in the final chapter just how odd some planetary systems and their denizens can be. For now, we simply note that this new understanding of solar systems in general opens up new possibilities for explaining *our* Solar System in particular. Clearly, ours appears a far more orderly place than many; possibly even most. But that apparent orderliness of middle age does not necessarily imply an entirely quiet youth. That was the mistake made by earlier theorists.

Although not universally agreed upon, one model of the Solar System widely accepted today is known as the Nice model (that's "Nice"—rhyming with "fleece"—as in the name of the French city where the model was first put forward, not "nice" as the opposite of "nasty"!). *Very* broadly and *very* briefly, the model may be summarized like this:

The four giant planets initially formed in nearly circular orbits a lot closer to one another than we see them now; from not far beyond the present orbit of Jupiter to beyond that of Saturn, though still well inside the contemporary orbit of Uranus. Beyond these planets millions of small icy bodies—planetesimals—orbited. For convenience, these may be called "comets" although many were considerably larger than the bodies to which this term is normally applied. Pluto-sized objects as well as bodies like the Earth-sized impactor believed to have toppled Uranus onto its side were included in their number, albeit together with multitudes of objects similar to the familiar comets that we know today. Being close to this belt of cometary bodies, Uranus and Neptune frequently encountered some of its members, deflecting many of them inward. Each encounter resulted in a loss of angular momentum for the comet and a gain for the planet and, although each contribution was negligible on its own, the sheer number of encounters meant that both planets gained angular momentum and slowly migrated outward. Comets deflected inward encountered Saturn, with similar results. On the other hand, Jupiter was then, as now, so massive that most bodies reaching its vicinity were deflected outward, ending up either in the distant Oort Cloud (more about this in the following chapter) or in interstellar space, expelled from the Solar System forever. The result was that Jupiter lost some of its angular momentum and migrated inward, although (fortunately for us!) halting its progress well short of joining the ranks of the hot Jupiters that we will meet in Chap. 6.

This slow dynamical evolution eventually brought Jupiter and Saturn into a 1:2 resonance, i.e. each planet moved into an orbit such that one complete revolution of Saturn coincided with two complete revolutions of Jupiter. This meant that the two planets returned to the same configuration after two revolutions of Jupiter and one of Saturn. Once this resonance was reached, things began to happen more speedily. Jupiter shifted Saturn even further outward and this in turn sent Uranus and Neptune into distant orbits of higher eccentricity, where they periodically penetrated into the disk of comets that began this whole process in the first place. These two intruding giants hurled comets every which-way. Many were thrown into the inner Solar System where they slammed into the Earth and the other rocky bodies in what is known as the Late Heavy Bombardment. Others retreated outward to form the Kuiper Belt. In the process, multitudes of encounters with these bodies reduced the eccentricity of the orbits of Uranus and Neptune (though not their average distances from the Sun) until, having cleared most of the smaller objects from their vicinity, they finally settled down into the orbits in which we find them today (some versions of the Nice model, however, have the two planets swapping places, such that Uranus was originally the most remote planet, but that remains an unsettled question and need not concern us here).

An ironic consequence of this model, or any recognizable variation thereof, is the unexpectedly important role played by minor bodies in the evolution of a planetary system. Traditionally, minor bodies such as comets and asteroids have been dismissed as almost unworthy of study, except for a small band of astronomers who found themselves drawn to this rather unfashionable subject. Percival Lowell, for example, dismissed comets as "bagfuls of nothing" and Patrick Moore as "[sometimes] striking, but … not very important" while Gauss, as we shall see in the following chapter, disparagingly described asteroids as "clods of dirt" while the expression "vermin of the skies" was heard on the lips of other astronomers in reference to both asteroids and comets. But if the Nice model is correct, the fate of planetary systems depends to a large degree on the number of comets and similar minor bodies whose formation precedes that of the planets themselves and of which those very planets are ultimately formed.

The Nice model, as it stands, is not perfect. For instance, the detailed structure of the Kuiper Belt has not been reproduced by any version yet put forward, but it does account for most of the features of the Sun's family and the correct model of this system will likely fall within its overall parameters. In particular, the presence and positions of the ice giants is accounted for in a way that earlier models could not accomplish and the broad structure of the Solar System itself explained in terms of various processes that appear to have sculptured every planetary system discovered to date. Only after finding other systems of worlds and discovering that they were far from being clones of our own, did the way open for models such as this and any real hope of a satisfactory account of the evolution of the Sun's planetary family made possible. It was the very weirdness of these alien systems that provided the key to unlock a greater understanding of our own!

3. Asteroids, Dwarf Planets and Other Minor Bodies

Back in 1988, Clifford Cunningham opened his book *Introduction to Asteroids* with the following words;

> Most amateur astronomers have never seen one; most professional astronomers hope they never have to see one. But these objects, variously known as planetoids, minor planets or asteroids, are increasingly being viewed as key to our understanding of the solar system.

I'm not sure if the remarks about amateur and professional astronomers still holds true (it certainly does not in certain quarters) but I doubt if anyone today would wish to argue with Cunningham's final phrase. From being mere specks of light in a telescope, asteroids have become worlds in their own right as visiting spacecraft beam back images of cratered terrain, boulder fields and, in some ways most startling of all, asteroid satellites. Indeed, it seems that the majority of moons in our solar system orbit, not the major planets, but asteroids and other similar minor bodies.

The name "asteroid" is not an especially good one. Originally coined by none other than W. Herschel, it essentially means "star-like" which, although giving a good description of the appearance of one of these objects in a telescope's eyepiece, is about as far removed as it can be from describing their true nature. There is little similarity between a vast ball of incandescent gas and a flying mountain of rock! For this reason, it was suggested that the early term "asteroid" be replaced by "planetoid" meaning something that resembles a planet. But the purists bucked about this as well, pointing out that these objects are not simply *like* planets, they *are* planets, albeit very diminutive ones. So the term "minor planet" was introduced to try to keep everyone happy. But further nomenclature problems arose more recently with the pesky problem of

Pluto. Demoted, with many howls of protest, from the ranks of the major planets, but held to be still a little too "major" to be written off as a minor planet, a new class of "dwarf planet" was recognized and, as we shall see, the parameters set for this class means that one of the traditional asteroids also has to be included within the new category. So not all the asteroids belong to a single category! In fact, there were problems here even before dwarf planets came onto the scene. Constant advances in telescope technology meant that smaller and smaller asteroids were being discovered and the smallest of these seemed to be stretching the "planet" definition a little too fine. It might not seem too bad to call an object a couple of 100 miles in diameter a minor *planet*, but what about a flying boulder 50 yards across?! We will not, however, enter into any disputes about names in these pages and will simply use the word "asteroid" without fear or favor.

The present time may or may not be the right season for discussing asteroids, as I suspect that many ideas about at least some of these bodies are about to change. As I write, the NASA probe *Dawn* is busy orbiting the brightest of these bodies—4 Vesta—beaming back to Earth unprecedented images of its surface. From there, if all goes well (to once more repeat the familiar Space Program caveat!) it will move on to 1 Ceres and repeat the task at that location. In the course of the following pages, some speculative ideas about Ceres will be aired but not long after you read them, data unavailable to the writer at this moment will (hopefully) start flowing in from the *Dawn* mission. This may confirm or overturn much of what will be said here. That is part of the risk and excitement of these interesting astronomical times.

Before turning our attention to some specific asteroids, a few words about these bodies in general will help set the scene.

Although asteroids may have occasionally wandered into the eyepiece fields of earlier astronomers and Vesta might even have been spotted by the naked eyes of one or two especially keen sighted ancients, the first person to officially discover an asteroid was Giuseppe Piazzi who found Ceres on January 1, 1801, during the course of a survey of stars that was to find maturity as the great Palermo Star Catalogue. This catalog ended up listing over 7,600 stars.

Piazzi at first thought that his moving star was a comet, rather as Herschel had mistaken the true nature of Uranus. However, when the "comet" showed no tendency to develop a fuzzy head and tail, he and other astronomers concluded (with Herschel's experience in mind?) that it must be a planet. Calculation of its orbit showed that it moved within the vast gap between the orbits of Mars and Jupiter and as such appeared to fulfill the expectation of many that a previously unknown planet should orbit somewhere in this region. The newly discovered world was given the name of Ceres, after the goddess of the harvest; a departure from the normally masculine mythological names hitherto given to most planets.

But what a puny little planet it turned out to be! Early estimates ranged from 1,579 miles (2,526 km) as determined by J. Schroeter to just 162 miles (259 km) as figured by Herschel. The true value is in between these extremes; some 582.5 miles or 932 km—tiny by planetary standards!

Some astronomers doubted that Ceres was alone, suspecting it to be just the tip of an iceberg. Possibly a very large iceberg. A discovery on March 28, 1802 suggested that they might be right. On that night, astronomer H. W. Olbers was observing Ceres when his attention was drawn to a similar moving star-like object. Forewarned by the knowledge of at least one small planet, this second object was quickly announced as another of the same variety. It is now known as 2 Pallas.

Some scientists seemed less than impressed by these objects. For instance, the mathematician K. Gauss, whose successful calculations of the orbit of Ceres enabled astronomers to continuing following it, nevertheless referred rather disparagingly to these two objects as "a couple of clods of dirt which we call planets"!

Just over 2 years later, on September 1, 1804, a third "clod of dirt" was discovered by Harding and on March 29, 1807, a fourth was spotted, once again by Olbers. These two came to be knows as 3 Juno and 4 Vesta respectively. The latter has the distinction of being the only asteroid which regularly becomes visible with the naked eye and at very favorable oppositions can be quite clearly—if faintly—seen without optical aid if one knows just where to look. Schroeter found that he could find it without optical aid soon after its discovery and is generally accredited as being the first person to see an asteroid with the naked eye. Nevertheless, as ancient Chinese astronomers included some stars of approximately the same brightness in their charts, it is possible that this asteroid had been seen by the ancients but not recognized as anything other than just another faint star amidst the multitude. Schroeter, by contrast, was well aware of what he was seeing.

Olbers' discovery of Vesta was not, in his opinion, due to pure chance. Following his earlier discovery of Pallas, he formulated the hypothesis which was to be resurrected several times in popular thought and science fiction as well as in more serious contexts. Namely, he speculated that the asteroids were fragments of a genuine full-sized planet that had for some reason "burst asunder". Consequently, even though fragments would be scattered and orbit at various inclinations to the ecliptic plane, the fact that they all originally diverged from the same point (the place where the hypothetical planet blew apart) implied that each should pass through two common points, namely, the orbital nodes. Olbers calculated that these points should lie in the constellations of Virgo and the western part of Cetus and it became his practice to search these regions each month whenever possible. Taking care to memorize the stars in the search areas, he performed the same exacting task repeated by visual hunters of novae, such as G. E. D. Alcock over 150 years later, namely, looking for a star which disrupted a memorized pattern. He believed that his discovery of Vesta vindicated his method and, by implication, the hypothesis upon which it was based. Nevertheless, the disrupted planet theory is no longer accepted by most experts in the field except in a *very* modified form as an explanation for "families" of asteroids. Most of these appear to have been caused by collisional disruptions of large parent asteroids at various times in the history of the Solar System. The alternative hypothesis for asteroids in general, put forward by J. Huth in 1807, seems closer to the truth, namely (to quote Olbers' account of his opponent's idea) "that the matter which formed the planets had coagulated into many small spheres in the space between Mars and Jupiter." It seems therefore, that Olbers' success in finding his second asteroid was due more to his diligent searching near the ecliptic than to the validity of his hypothesis. Incidentally, he did not stop searching after Vesta was found, but kept going until 1816 with, however, no further asteroid discoveries.

Indeed, no more asteroids were found until 1845 when Driesen postmaster K. Hencke discovered 5 Astraea. By 1852, 21 were known, many being picked up by amateur astronomers. Numbers kept increasing to such an extent that some astronomers were beginning to think that keeping track of them was more trouble than it was worth, deflecting time and resources from what they considered to be more important tasks. Phrases like "a plague of asteroids" and "vermin of the skies" were coming into vogue in certain quarters as the number of listed asteroids grew to 300 by the last decade of the nineteenth century. Then the numbers *really* started growing!

This explosion in discoveries came about by applying the new technology of astro-photography to the search. The pioneer of this method was Maximilian Wolf, director of Konigstuhl Observatory, Heidelberg. Wolf employed portrait lenses of 6 in. (15 cm) diameter and focal lengths of 25 in. (63 cm) and 30 in. (76 cm). Using these two lenses, he exposed duplicate plates; plate A being initially exposed for 1 h and plate B for another hour. Then, plate A was exposed again for a further hour. If an asteroid happened to be moving through the field, part of the space occupied by its trail on A would be vacant on B and vice versa. Wolf then laid the two plates on a retouching frame, one on top of the other, and examined them through a magnifying glass. By introducing this technique from 1891, Wolf discovered a grand total of 231 asteroids.

Others soon began emulating Wolf's method—and his success. In 1898 a surprising discovery was made by G. Witt, director of the Urania Observatory in Berlin. Prior to that year, all the discovered asteroids were confined to the broad region of space between the orbits of Mars and Jupiter. They were, in the terminology of later times, "main-belt asteroids". But Witt's body turned out to have an orbit with perihelion well inside that of Mars! The first Mars-crossing asteroid (now known as 433 Eros) had been found. Then on February 22, 1906, Wolf found a distant asteroid which turned out to be moving in the same orbit as Jupiter, but preceding the giant planet by 55.5°. Lund Observatory's C. V. Charlier recognized that the position of this object (now known as 588 Achilles) coincided with one of the so-called Lagrangian points associated with Jupiter's orbit. These "points" are five positions in an orbital configuration where a small object affected by gravity alone can theoretically remain stationary relative to two larger objects. In the case of Achilles, the two larger objects are Jupiter and the Sun.

Before 1906 was over, a second similar asteroid was found *leading* Jupiter by 55.5°. This object became known as 617 Patroclus (by the way, notice how fast the number designations were growing by 1906 as more and more asteroids came to be recorded!). These were the first of a large group that became known as *Trojan Asteroids* or, simply, *Trojans*. The number of Trojans larger than 1 km is estimated to be around one million. Incidentally, that is also the estimate for similar sized asteroids in the Main belt, so what these little objects lack in size they certainly compensate for in number!

Strictly speaking, asteroids like Achilles and Patroclus should be called "Jupiter Trojans" as objects at similar points (the Langrangian L4 and L5 points, 55.5° ahead or behind the major planet) have been found associated with other planets as well. Neptune has some, a couple have been found associated with Mars and in 2010 the space-based infrared observatory *WISE* found a small one associated with Earth. This Earth Trojan—which has been given the provisional designation of 2010 TK7—is just 1,000 ft (300 m) in diameter.

Two discoveries that were later to give the search for asteroids a new and, as some may interpret it, more ominous twist took place in 1932. March of that year saw the discovery of 1221 Amor and the very next month brought to light 1862 Apollo. The orbits of both of these asteroids crossed that of Earth, for the first time hinting at the possible existence of large bodies potentially able to strike our planet. The increasing number of the Amor-Apollo class of asteroids today drives several sky patrols using robotic telescopes and CCDs to try to assess the number of near-Earth asteroids posing a potential threat to our planet and (hopefully) finding any such threat sufficiently in advance to deflect the offending body or deal with it in some other suitable manner. Thus far, no sure threat has been uncovered, but the number of asteroids being found by these surveys would have surely blown the minds of asteroid pioneers like Olbers and even Wolf. By the way, the distinction between true Apollos and Amors is somewhat blurred, but technically an asteroid with perihelion slightly outside Earth's orbit is an Amor and one that ventures within, an Apollo. Both can come close to Earth, and some Asteroids move in orbits that oscillate from one to the other, further fudging the boundary.

A curious feature of the Amors is worth mentioning here. Asteroids normally move across our skies from west to east against the background stars. They are said to have *direct* motion, just like the outer planets. However, in common with the outer planets they can, around the time of opposition, be temporally overtaken by Earth (moving with higher relative velocity) and for a while appear to move backward. i.e. westward relative to the stars. This *retrograde* motion is only temporary, but is displayed by Amor asteroids as well as by Mars, Jupiter and all of the outer planets. Nevertheless, Amor asteroids are also capable of coming very close to Earth if the times of opposition and perihelion nearly coincide. On these occasions, an Amor's rate of movement across our skies will be very rapid; so rapid in fact that the usual retrograde loop will not occur and the asteroid's direct motion remains uninterrupted. This has the strange consequence of producing three oppositions during a single close approach! At more remote oppositions however, this effect does not occur and the usual retrograde loop is faithfully followed by Amor asteroids.

The variety of asteroid orbits now known would also have amazed Olbers and colleagues. In 1949, an asteroid (designated 1566 and suitably named Icarus) was discovered that passed well within the orbit of Mercury, approaching to just 18 million miles (28 million kilometers) of the Sun. Although this small perihelion distance stood as a record for many years, a number of other "sun-approachers" are now known. In 1983 the *IRAS* infrared survey satellite discovered a small asteroid provisionally designated as 1983 TB and since given the permanent designation and name of 3200 Phaethon. This curious object swoops passed the Sun at just 13 million miles (21 million kilometers) and also has a remarkably high eccentricity of 0.8979. An even bigger surprise came when Fred Whipple noticed that its orbit is remarkably close to that of the annual Geminid meteor shower, making this the first asteroid to be identified as the parent of a meteor stream. Subsequently other asteroids of remarkably small perihelion distance have been found; one coming to approximately two million miles (or three million kilometers) of the Sun—a very small distance. This one becomes a very hot rock indeed!

At the other end of the scale, we know of asteroid 944 Hidalgo whose eccentric orbit takes it from just beyond that of Mars to out near Saturn, of the Aten group of asteroids (named for 2062 Aten, discovered in 1976) whose orbital semi-major axes are smaller than Earth's and whose orbits overlap that of Earth near their aphelia, asteroid-like bodies orbiting beyond Pluto and even some Earth-approaching bodies which can become temporary satellites of our planet. With respect to this last, a tiny little asteroid—no larger than a small cottage—designated 2006 RH120 became a second moon of Earth between September 2006 and June 2007 before escaping into its heliocentric orbit once more!

Familial Asteroids and Gaps in the Belt

Turning our attention back to the main belt, we note that this can no longer be thought of as a simple and structureless band of rocky brothers. Back in 1918, Kiyotsuga Hirayama found that a significant proportion of main-belt asteroids are grouped together in orbital families. Initially, he found just three such groupings (*Hirayama families* as they are now designated); the Themis, Eos and Kronos families, each named for its best known member. Nowadays, we suspect that half of the asteroids of the belt belong to families. There are even sub-families; second generation associations within broader groupings such as, for example, the Beagle family within the larger Themis family. Hirayama families are thought to result from the disruption of a single large asteroid and the smaller sub-groupings presumably betray subsequent collisions, either between two family members or between a member of the group and an interloper asteroid.

As well as families of asteroids sharing similar orbits, the main belt also has regions where there are few, if any, observable objects. Percival Lowell likened these "gaps" to the divisions in Saturn's rings though, of course, on a vastly different scale. Their presence was first noted in 1857 (though not officially announced for another 9 years) by Daniel Kirkwood (1814–1895), after whom they are named. Kirkwood found that the gaps occurred where the orbital period of hypothetical particles were in resonance with the orbital period of Jupiter. The main Kirkwood gaps occur at mean orbital radii of 2.06, 2.5, 2.82, 2.95 and 3.27 AU. These correspond to resonances of 4:1 (i.e. four orbits by a hypothetical object at 2.06 AU from the Sun for every orbital revolution of Jupiter), 3:1, 5:2, 7:3 and 2:1. Narrower and/or weaker gaps also occur at the 9:2 resonance (1.9 AU), 7:2 (2.15 AU), 10:3 (2.33 AU), 8:3 (2.71 AU), 9:4 (3.03 AU), 11:5 (3.075 AU), 11:6 (3.47 AU) and 5:3 (3.7 AU). Kirkwood explained that "a planetary particle ... always coming into conjunction with [Jupiter] in the same part of its path" will "become more and more eccentric" ending up in unstable orbits that cause their eventual removal from the main belt. In 1982, Jack Wisdom used a set of 300 hypothetical "test asteroids" in the neighborhood of the 3:1 resonance (around 2.5 AU) to determine if this process could produce a Kirkwood gap, like the one actually present at that location, within a time frame of two million years. The answer was a clear positive. Asteroids were removed from the gap, 84 becoming Mars crossing within 300,000 years and 88 achieving orbital eccentricities as high as 0.3 within a million years. Moreover, a 1986 study by Carl Murray found that asteroids near the center of the 2:1 resonance (around 3.27 AU) will eventually end up in Jupiter-crossing orbits.

Asteroids ejected from the Kirkwood gap regions are believed to supply at least some of the Mars-crossing and Earth-crossing bodies that we met earlier, although there is still debate as to what percentage of this asteroid population may be comprised of bodies coming in from the outer Solar System or might even by inactive comets; either "dead" short-period comets that have run out of volatiles or "sleeping" ones whose volatiles have become insulated by a refractory layer of carbonaceous compounds or a mantle of siliceous particles. Nevertheless, despite a probable cometary component amongst the Earth-crossing asteroids, the existence of resonances between Jupiter and asteroids within the main belt has surely contributed material to the region of the terrestrial planets and directly to the surface of Earth itself.

Light and Dark Asteroids

Broadly speaking, asteroids may be divided into dark bodies with the low reflectivity of carbonaceous meteorites and relatively light colored objects with a reflectivity closer to that of the most common type of meteorite; the stony "ordinary chondrite". Objects belonging to the first general class are designated C-type asteroids and the second, S-type. Needless to say, these broad types have become increasingly subdivided into sub-classifications as more detailed information about the reflectivity and surface properties of asteroids have been discovered. Nevertheless, this broad division will suffice for our purposes.

Asteroids of the C-type are the most numerous and strongly dominate the outer main belt around 3 AU from the Sun. The largest main-belt asteroid of all—Ceres—belongs to one of the subspecies of this class. Other close relatives of the C-types (some of the sub-classes of this group) dominate even further out, with reddish bodies of low reflectivity (once known as RD but now simply as D-type) comprising most of the asteroids around 5 AU, i.e. in the region of Jupiter's orbit. Jupiter's Trojans are also of this type. Toward the inner edge of the main belt, around 1.5–2.0 AU or just beyond Mars, S-types dominate. C-types are a small minority within 2.5 AU, effectively vanishing from the population of main-belt asteroids within 2 AU, while S-types pretty much disappear beyond 3.5 AU. The Earth-approaching population of asteroids includes some that are reflective (S-type and similar) and others that are dark and of low reflectivity (C-type and related). Asteroids located beyond the orbit of Jupiter are of the darker variety.

Dawn of Greater Knowledge of Two Asteroids

Vesta

A new goal of space exploration was reached on July 16, 2011, when NASA's *DAWN* spacecraft became the first to go into orbit around a main-belt asteroid; 4 Vesta. Having a diameter of 331 miles (530 km), Vesta is the second largest of the asteroids of the main belt, and accounts for almost 10 % of the belt's mass. Its diameter is, nevertheless, significantly less than that of Ceres (583 miles or 932 km). But although Vesta is neither the largest nor the closest of the inhabitants of the main belt, it is distinguished by being the brightest and at a good opposition is relatively easily spied with the naked eye if one has a suitably dark sky and knows just where to look. At less favorable oppositions it needs opera glasses unless one is exceptionally keen sighted and blessed with very good rural skies. Its comparative brightness is the result of a surface that is very reflective by asteroidal standards, courtesy of a composition similar to the volcanic rocks and minerals found in the crusts of the planets Earth, Venus and Mars, as well as on Earth's Moon. These minerals include the likes of the common iron-and-magnesium-bearing silicate olivine, found in volcanic rocks that formed deep within Earth's crust. Its appearance on Vesta suggests volcanic activity not unlike that witnessed on Earth.

The asteroid's spectrum has long been known to match that of a class of closely related meteorites known collectively as HED's ("Howardites, Eucrites and Diogenites"), all of which give evidence of having been formed within the mantle or crust of a relatively large parent body before being violently ejected into space as the result of a large and violent impact. If these meteorites really did originate at Vesta—and their association with the asteroid appears strong—they tell us something of the early history of this body. Because of their similarity to volcanic rocks, HED meteorites inform us that Vesta must have differentiated at some time in the remote past. Radioactive elements, and maybe other heat sources as well, apparently raised the asteroid's internal temperature high enough for it to become molten, segregating the body into core, mantle and crust like a miniature version of the Earth. This is why some planetary scientists have even gone so far as to call Vesta the "smallest terrestrial planet". Actually, because it is too small even to achieve hydrostatic equilibrium, it cannot really be classed as a dwarf planet (more will be said about this topic later), so calling it the smallest terrestrial planet is something of an exaggeration. Yet, the reason for doing so is clear enough. Vesta appears to be the smallest known object that has been differentiated in the way that true terrestrial planets are and to that degree looks like a very, *very*, small counterpart of Planet Earth! Its density of 3.4 times that of water is a little more than that of the Moon (Fig. 3.1).

Vesta is slightly oval in shape. It apparently is not quite massive enough to have settled into the approximately spherical shape betraying hydrostatic equilibrium.

For the first time, interesting landforms have been revealed on its surface by *DAWN's* cameras. As expected for an ancient object

Fig. 3.1 Comparison sizes of several asteroids (Credit: NASA)

devoid of protective atmosphere, the surface of Vesta is heavily cratered. One conspicuous triple crater, affectionately named "The Snowman" because of its appearance, may mark the impact sight of a binary asteroid. The two largest craters making the "body" of the "snowman" merge together without a dividing rim and give the appearance of having been formed simultaneously. The largest of the pair is some 41 miles (65 km) across. The "snowman's" "head" (a small fully formed crater almost touching the other two) may have resulted from a different impact at another time; its position so close to the others being simply fortuitous.

The real oddity however is a colossal crater at the south pole of the asteroid; a crater so vast that it was first observed, not by *DAWN* but by the *Hubble Space Telescope*. *DAWN's* images however, reveal its true complexity. At its center, a mountain towers over 11 miles (18 km) above the crater floor. That floor itself lies between 8 and 9 miles (between 13 and 15 km) below the mean surface of the asteroid, implying that the mountain summit stands between 2 and 3 miles (3 or 5 km) above the mean surface. Surrounding the crater is a series of grooves extending all the way the equator. Some of these are around 6 miles (10 km) wide. Huge cliffs, compared by planetary scientists to the ice cliffs of Saturn's moon Miranda or even the majestic canyon walls of Mars' *Valles Marineris*, terminate some of these grooves. Clearly, little Vesta harbors spectacular and dramatic scenery, but the full extent of this is probably not realized until it is scaled up to terrestrial levels. If Vesta's 331-mile diameter is expanded to Earth's 7,900 and the huge south polar mountain increased accordingly, an equivalent peak on our planet would tower a fantastic 263 miles above the floor of a crater some 203 miles below sea level. The peak's summit would be some 60 miles above sea level! That would be, for all practical purposes, on the edge of space. Someone standing on the summit would see larger meteors burning out at eye level and others passing below—*well* below—one's feet!

The south polar crater of Vesta is thought to be where the fragments that eventually arrived at Earth as HED meteorites were blasted out of the asteroid. Almost certainly, it is also the point of origin of the larger fragments that now orbit as individual asteroids in the Vesta family. Yes, Vesta has her children, despite being named after the virgin goddess of the Earth.

As these words are being written, *DAWN*'s examination of Vesta is continuing and we can look forward to more information—and maybe some surprises—in the near future. Then, in July 2012, the spacecraft is due to leave orbit of Vesta and head for its rendezvous with 1 Ceres slated for February 2015. If all goes according to plan, by the time you read these words, it will be well on its way.

Ceres

At the time of writing, Ceres remains unvisited by *DAWN*, but hopes are high that the spacecraft's performance at Vesta will be repeated there come 2015. Everything said here about Ceres must, therefore, be more or less tentative but much of it will hopefully be either confirmed or overturned not too long after this book is available (Fig. 3.2).

Fig. 3.2 The dwarf planet Ceres as imaged by the *Hubble Space Telescope* (Credit: NASA/ESA)

Ceres is the largest of the asteroids and, as we shall see later, the only one that is massive enough to be in hydrostatic equilibrium and, therefore, deemed a genuine dwarf planet. Unlike the other asteroids, it is approximately spherical in shape, and contains approximately one third of the mass of the main belt. Even so, it still contains just 0.0128 times the mass of our Moon and a mere 0.00015 that of Earth. Cosmically speaking, it is hardly a massive object! (Fig. 3.3)

Spectroscopic studies have determined that Ceres has a surface consisting of dark materials similar to those comprising carbonaceous chondrite meteorites. It is classified as a G-type asteroid, one of the sub-groupings of the populous C-type. Unlike most other asteroids, its spectrum indicates the presence of hydrated minerals on the surface, together with iron-rich clays and minerals such as dolomite and siderite. The presence of hydrated materials is seen as supporting the view that Ceres was formed from a mixture of ice and rock and that a good deal of water—in the form of ice—may still be present within the asteroid.

Asteroids, Dwarf Planets and Other Minor Bodies   135

Fig. 3.3 Comparison sizes of Earth, Moon and Ceres (Credit: NASA)

There are other reasons for thinking this as well. Spectral features suggestive of surface frost have been noted and the *International Ultraviolet Explorer* satellite recorded statistically significant concentrations of hydroxyl ions in the region of Ceres, indicating a very tenuous atmosphere or extremely weak ultraviolet coma. With a semi-major axis of 2.8 AU, Ceres is too warm for frost to be stable on its surface and the most straightforward explanation for both of these observations is the slow migration of water from deep within the body forming, first of all, a layer of surface frost which in turn gradually sublimates away into space. This sublimating ice is continually being replaced by more water seeping up from below the surface. Ultraviolet radiation from the Sun breaks down the molecules of water vapor, giving rise to the thin cloud of hydroxyl ions apparently observed by the *IUE*.

Planetary scientists are not, however, agreed as to what really does lurk beneath the surface of Ceres. One model suggests a differentiated object with a rocky core overlain by an icy mantle about 60 miles (around 100 km) thick and comprising between 23 % and 28 % of the asteroid's total mass and around 50 % of its volume. This ice mantle in its turn is overlain by a thin crust of dusty material having a similar composition to that of carbonaceous meteorites. According to this model, Ceres contains more water than the fresh water stocks of Earth and there is speculation that some of this may have been—or may *still* be—in the form of an underground ocean.

An alternate model suggests that Ceres remains undifferentiated (or, at most, partially differentiated) and that the interior of the asteroid is quite porous. Supporters of this model point out that the penetration of rock into ice layers should form salt deposits, for whose existence no evidence exists. This may imply that Ceres has no large shell of ice, but consists entirely of low-density hydrated material. Hopefully, *DAWN* will send back information in 2015 that should help decide which (if either!) of these models is nearer to the truth.

Is There *Life* on Ceres!?

The prospect of an underground ocean on Ceres, possible on the first of the above models, is an interesting one and has led some people to speculate about the possibility of Cererian *life*! The existence of life on an asteroid or, to be more accurate, *in* an asteroid, is something not often considered, although it is raised from time to time in the context of certain structures found within carbonaceous meteorites and controversially identified as extraterrestrial microfossils. Some adventurous folk have even speculated that Ceres may have seeded Earth with life. They argue that *if* Ceres has (or once had) an underground ocean and *if* simple life forms found their home there, violent collisions with other bodies in the asteroid belt may have launched life-bearing fragments into space, some of which eventually made it to Earth with the seeds of life still viable. The low gravity of the asteroid would make it far easier for a fragment to become "spaced" than, say, a piece of Mars or even the Moon. Yet, meteorites of Martian and Lunar origin have been identified, so Cererian meteorites are at least a possibility.

The main problem with this hypothesis—quite apart from the controversial existence of either oceans or life on Ceres—is the evidence that Ceres has not been extensively battered during its lifetime. Hubble telescope images do show several dark spots which are probably impact craters, plus a bright feature of unknown nature, but nothing to be compared with, say, the polar crater of Vesta. Small fragments of the asteroid most probably are floating around somewhere in space, but they are not likely to be very numerous and if any have reached our planet, they must have been few and far between. That does not disprove the interesting suggestion of Cererian life colonizing Earth, but it does make it less likely. (Viable spore blown out from a "cometary" jet on Ceres is another possibility, but until more is known about the asteroid any wild speculation of this type had, perhaps, better be confined to private conversions between friends.)

Apparently confirming the lack of major impacts is the lack of a Cererian Hirayama family. Unlike Vesta, Ceres is a loner amongst the asteroids. As a matter of historical interest, Ceres was once thought to have a family, but spectroscopic analysis discovered that the presumed family members have a different composition to Ceres and that the apparent orbital similarities are simply coincidences—Ceres is a gate-crasher in another asteroid's family! Consequently, what used to be called the Cererian family has been renamed the Gefion family after its most prominent genuine member 1272 Gefion.

Ceres is most probably the one remaining protoplanet in the asteroid belt; the one that avoided being shattered, although there is a 10 % probability that it may actually have originated in the Kuiper belt beyond the orbit of Neptune. The first (and greatest) probability is the favorite one however. If that is correct, Ceres gives us a look at the last remaining representative of the objects that became the progenitors of today's asteroid population. As such, it is something of a cosmic fossil left over from an ancient era of the Solar System; an era before the planetary system as we know it today was formed.

One last thing about Ceres. It is a place in the Solar System where—on rare occasions—solar transits of all the rocky planets (Mercury, Venus, Earth and Mars) may be witnessed. Mercury provides the least spectacular but most frequent transits, crossing the Sun's disk every few years. Its most recent were in 2006 and 2010. Venus crossed the Sun as seen from Ceres most recently in 1953 and will do so again in 2051. One wonders if that will be witnessed by a surface probe or (just maybe if the enthusiasm for space exploration picks up again) by an astronaut! Earth last crossed the face of the Sun in 1814 and will repeat the performance in 2081, but

Mars gives potential Cererian astronomers a longer wait. It has not transited the Sun since the year 767 and will not give another performance until 2684!

With (hopefully!) the first probe to Ceres so close, we hope that much more will become known about this miniature planet in the near future. No doubt, things that we now think we know will be changed and some surprises will be in store. One thing is for sure though. Ceres will emerge as a fascinating little world. The days when it was dismissed as a "clod of dirt" or numbered amongst the "vermin of the skies" are—thankfully—long gone!

Carbonaceous Meteorites: Pieces of Asteroids?

The similarity in color between C-type asteroids and carbonaceous chondrite meteorites is quite striking and we may wonder if objects with compositions similar to that of Ceres provide the sources of these objects. In other words, do the sisters of Ceres have fragments of themselves in our laboratories and museums? This question is all the more intriguing in view of the biologically important organic compounds found in these meteorites and the controversial hint that some of the carbonaceous structures might even constitute something approaching simple organisms. Parent objects of these meteorites might have an important story to tell us about life itself!

This latter fits well with speculation as to the possible biological significance of Ceres but, as already stated, it is doubtful if carbonaceous meteorites really do hail from that source given the apparently undisturbed nature of this dwarf planet. They are more likely to have come from disrupted objects of similar composition. Asteroid 19 Fortuna has a very similar appearance to carbonaceous meteorites of CM2 type and CR2s look to be good matches to the Geminid asteroid Phaethon. But what of the most primitive and fragile CI1 types? They are the ones that have the largest organic component and they are also the ones with the highest content of water, chemically combined with various minerals within their matrix. It would be very interesting to find out where these interesting meteorites originate.

Asteroids, Dwarf Planets and Other Minor Bodies   139

There has long been a strong suspicion in some quarters—though equally strongly criticized in others—that these meteorites are pieces of comets. Unfortunately, most of the meteors whose cometary associations are beyond serious doubt enter our atmosphere at such high velocities that they stand no chance of reaching the ground as anything more substantial than a scattering of microscopic dust particles. The chief exception is probably the Taurid stream, a broad debris trail left in the wake of the short-period Comet Encke and giving rise to a weak but extended sprinkle of meteors from September 17 until December 2, with greatest activity in early November. These meteors of the northern fall are seen as Earth meets the particles on their way toward perihelion. However, we also cross this stream as the particles move outward from perihelion, giving rise to another weak shower of meteors from June 5 to July 18, reaching a relatively flat maximum around June 30. Co-incidentally, this shower also radiates from Taurus and is known as the Beta Taurids. Because the radiant point is quite close to the Sun in the sky, these meteors fall in daylight and are observed by radar (Fig. 3.4).

The Taurid/Beta Taurid complex is interesting in that it harbors some pretty large objects capable of producing brilliant fireballs which can penetrate to quite low altitudes in Earth's atmosphere. They are also a lot slower than most cometary meteors and have a higher tensile strength than much fragile comet debris. If any meteor shower derived from a comet drops meteorites, chances are it will be the Taurids.

From a study of bright Taurids by the Meteor Section of the British Astronomical Association, published by Section Director K. B. Hindley in 1972, a linear relationship was found between the start height and end height of Taurid fireballs in the magnitude range of –2 to –9. The end height dropped rapidly with increased brightness. Fireballs brighter than about –7 should penetrate deeply enough into the atmosphere to produce sonic booms, which have indeed been recorded as accompanying the brighter Taurids. Extending the linear association beyond the actual data plot, this study suggests a dropping of the end height to just 18–19 miles (about 30 km) for Taurids of –10 or –11 magnitude and to just 15–16 miles (about 25 km) for those of –12. End heights below 18 miles are associated with potential meteorite-dropping

Fig. 3.4 Artist's impression of the meteorite that struck the Moon on 7 November 2005. This was probably a Taurid in which case it would have been travelling at 17 miles per second, measured about 6 in. in diameter and exploded with the force of 32 lb of TNT (Credit: NASA/MSFC)

fireballs. Although the average fireball brightness in the study was a relatively modest −4.8, one object reached about −12, another −11 and yet another, −10, with three in the −9 region. According to the results of this analysis, some of these may well have dropped small meteorites.

Yet all is not quite so straightforward. On October 19, 2010, a Taurid fireball reaching a brightness of −12 lit up the Spanish skies and was recorded on the cameras of the Spanish meteor network. Remarkably, this object faded out at an altitude of 40 miles (64 km); far too high for anything more than thinly distributed microscopic dust particles to reach the ground. Moreover, the fireball suffered one flare along its path, presumably caused by splitting of the meteoroid; far from a common feature of Taurid meteors and (like the end height) indicative of a lower tensile strength than is usual for members of this stream. Clearly, there

Asteroids, Dwarf Planets and Other Minor Bodies   141

is quite a variation amongst Taurids, which may modify Hindley's conclusions somewhat.

Another fireball study, this time by G. W. Wetherill and D. O. ReVelle and published in 1982, found that Taurid fireballs do indeed differ amongst themselves in terms of tensile strength. Using data from the Prairie Network, they found that bright Taurid meteors (which formed only a relatively small percentage of their study of the general fireball population) could be divided into "core" members of the stream and "outliers". Core members tended to have lower strength and even bright ones burnt out at altitudes too high to drop meteorites. Outliers, on the other hand, appeared stronger on average and the biggest and strongest of these had properties consistent with the weaker recovered meteorites such as CI1 chondrites. No actual meteorites could be identified with any of these fireballs, but the authors concluded that small and weak ones were possible from a minority of them. This conclusion is broadly in line with Hindley's except that the newer study implies that any Taurid meteorite will more likely come from an outlier than from a core member of the stream.

Wetherill and ReVelle put forward two suggestions as to why this might be so. First of all, they suggest that the stronger outliers might be interlopers; asteroidal fragments that just happen to have wound up in Taurid-like orbits. This suggestion they find unlikely, as the likelihood of asteroidal debris ending up in orbits of small aphelion and small perihelion (i.e. Taurid-like orbits) is small. Moreover, we might add that it would seem a whopping coincidence if the Taurid stream was encased within a broader and more diffuse system of objects having similar orbits but a totally different origin (Fig. 3.5).

The second hypothesis is much more likely. We know that meteor streams diffuse away over time thanks to gravitational perturbations and other disruptive effects which need not concern us here. Therefore eventually "shower" meteoroids will drift so far from their original paths as to lose any obvious relationship with their parent stream. In effect, they join the general field of "sporadic" meteoroids. According to the second hypothesis, the Taurid outliers are simply objects that have drifted from the main stream, but not so far as to join the sporadic background. Such a drift takes time and, although Taurids as strong as CI1 meteorites are presumably a very small component of the whole stream population, they have a better chance of surviving the impacts of small meteoroids and arriving intact in the outlier field than their more fragile siblings. Therefore, while any fireball-sized meteoroid within the core population will most probably by of the more fragile type, members of the outlier field have a far higher chance of being of the stronger variety; the biggest, toughest, brutes of outliers even being capable of dropping meteorites.

Unfortunately, no Taurid meteorite has been recognized and no CI1 fall has coincided with either Taurid or Beta Taurid activity. At least no *confirmed* fall! But there is one historic incident that might—just might—be associated with the stream; in particular, with the Beta Taurid branch in June/July.

The event to which I refer is the meteorite fall at Veliky Vstyug in Russia on July 3 (Gregorian calendar) 1290. Admittedly, this was a long time ago, but the Taurid stream was also active then and may have been stronger than it is today. The November Taurids (and presumably the daytime Beta Taurids as well) were indeed more numerous in the eleventh century and activity may still have been higher than at present as late as 1290. Be that as it may, that year the saint and hero who later became known as the Righteous Prokopiy prophesied that the community would be struck by fire, whirlwind and a rain of rocks. Naturally, the people were a little unnerved by this and hastily repented of their evil ways, in response to which Prokopiy prayed for their deliverance from the catastrophe which he foresaw happening. According to the story, fire, whirlwind and a rain of rocks from the sky did indeed occur, but the event missed the community and struck uninhabited land. To a modern reader, the event certainly sounds like the fall and airburst of a large meteorite and has always been interpreted that way by Russian scientists. Not only is the date consistent with a Beta Taurid association but there is also an indication that the meteorite arrived from the "direction of the Sun"; consistent with the trajectory of a Beta Taurid. The area of the fall quickly became a pilgrimage site and has remained so to this very day—Marxist revolution, atheist indoctrination and eventual capitalist expansion notwithstanding.

Meteorites have been incorporated into some Russian churches and given the religious significance of the abovementioned event, one would expect that fragments of the fallen rocks would have been collected by monks and pilgrims and placed in churches and shrines throughout the region. Yet no fragment of the meteorite has been identified. The Church of the Righteous Prokopiy now stands near the place of the fall, but contains no extraterrestrial material in its walls. Even the very dark colored rock on which Prokopiy is said to have sat, and which was long regarded as one of those that fell from the sky on that fateful day, has more recently been identified as entirely terrestrial.

In a perverse sort of way, the non-survival of meteorite samples tends to strengthen the suggestion that this may have been a CI1. These objects very quickly decay. A piece of the famous Orgueil meteorite that fell in 1864 was said to have "disintegrated and gave off a nasty smell" when dropped into a glass of water and another piece of the same object became soft like modeling clay when slightly wet. Samples of this and similar meteorites require special air-tight containers for preservation, protected from atmospheric humidity and other traces of moisture. Maybe pilgrims, priests and monks did indeed collect samples of the Veliky Vstyug meteorite, only to have them quickly turn to dust. It is interesting that the stone on which Prokopiy used to sit is dark in color. Maybe the saint originally did sit on a piece of the meteorite but, after it crumbled away, replaced it with a look-alike. Or, more probably I think, as pieces of the meteorite were being collected, a large lump of rock of similar color was genuinely mistaken for a piece and accordingly given pride of place.

Of course, all of this speculation finally *proves* nothing. All we can say is that the events at Veliky Vstyug on that long ago day *appear* to be consistent with the arrival of a large Beta Taurid meteorite and, further, that the little we know of the event *appears* consistent with it having been of CI1, or closely related, composition. But we cannot know whether this apparent agreement is anything more than a reflection of our ignorance as to what really took place on that day so long ago.

Another piece of evidence in favor of a cometary origin for CI1 meteorites was presented by Matthieu Gounelle, Pavel Spurny and Philip Bland in their in their 2005 study of the trajectory of the Orgueil meteorite. This famous object, which fell in France on May 14, 1864, remains the largest fall of CI1 material yet found.

(The Revelstoke CI1 of March 31, 1965, was probably a larger body when it entered the atmosphere, but little reached the ground and only a miniscule amount of material was recovered). By examining the numerous descriptions of the Orgueil fireball, Gounelle and his co-authors found that they could pin down the object's atmospheric trajectory sufficiently to enable a series of possible solutions for its pre-impact orbit to be derived. What most impressed these authors was that the computed orbits for the most realistic estimates of the meteorite's speed and trajectory all turned out to be very comet-like, with aphelia beyond the orbit of Jupiter. All the possible orbits also had remarkably low inclination; less than 1° in every instance and a mere 6 min (!) for the more probable solutions. According to Gounelle et al., these results suggested that Orgueil might be a fragment from a comet of the Jupiter family or even of one having a Halley-like period.

The biggest objection to this conclusion, in Gounelle's opinion, concerned the difference between the deuterium content in the meteorite and that of the comets whose deuterium had been determined at the time of writing, plus the widely held opinion that cometary material had probably not been exposed to liquid water in the manner of CI1 meteorites. Each of these possible objections has been weakened by further research since 2005.

First, at the time of Gounelle's paper, published information on the deuterium content of comets rested upon analysis of the three bright "H-comets" of the final two decades of the previous century; Halley, Hyakutake and Hale-Bopp. These were all rich in deuterium. However, in November 2010 the short-period comet 103P/Hartley was found to have a deuterium content exactly equal to that of meteorites such as Orgueil—and, for that matter, Earth's water. Hartley is a comet of the Jupiter family, so the deuterium content of this object supports (or at the very least presents no difficulty for) the suggestion that Orgueil might be a fragment of a Jupiter-family comet (Fig. 3.5).

Secondly, in 2004 the space probe *Stardust* flew through the dusty coma of another Jupiter-family comet, 81P/Wild, collecting a sample of the comet's particles and returning them to Earth 2 years later. To the surprise of many astronomers, some of this material proved to be quite "processed" and showed clear signs of aqueous alteration. Apparently, some cometary materials have

Asteroids, Dwarf Planets and Other Minor Bodies 145

FIG. 3.5 The active nucleus of Comet 103P/Hartley, 3 November 2010 imaged by *EPOX* 1 spacecraft (Credit: NASA)

been altered by water; another plus for the suggested cometary origin of Orgueil.

Yet, grounds for skepticism remain. For one thing, the time that a meteorite has spent as a free orbiting body in space can be ascertained by gauging the amount of cosmic ray "damage" to its surface material. In Orgueil's case, this "cosmic ray exposure age" (CRE) is around five million years. Yet, dynamical calculations show that the maximum time that a body can spend in a low-inclination Earth-and-Jupiter crossing orbit is of the order of 100,000 years. That is the greatest length of time that it can expect to avoid collision with Earth or Jupiter or such a close encounter with the latter planet as to be gravitationally sling-shot right out of the Solar System. As this dynamical lifetime is short by comparison with the CRE age it is pretty clear that the orbit on which the meteorite arrived at Earth in 1864 could not have been its original one and—ipso facto—not the orbit of its parent body. This does not mean that the original orbit of the meteorite and its parent was not that of a comet, but it does mean that it did not *need* to be that of a comet. Asteroidal fragments from near the resonance zones can evolve dynamically from typical main-belt orbits to Earth-crossing ones in the order of a million years. Close approaches to Earth can then pump up these orbits into Jupiter-crossing ones in a similar time span. Indeed, objects close to the 2:1 resonance can become Jupiter-crossing over these time periods and it is interesting to note that this resonance lies very near the outer limits of the Themis family of asteroids, whose spectral characteristics happen to appear quite close to those of CI1 meteorites. Some of the relatively few asteroids orbiting within the Kirkwood gap marking this resonance are actually thought to be prodigal members of the Themis family. Could it be that Orgueil and its brethren originate not from comets, but from Themis family asteroids?

This asteroid family has indeed been suspected of being the home of carbonaceous meteorites, yet recent studies have found that it could at best supply only a small percentage of the observed influx of these objects. The main problem is the eccentric orbits and consequent high velocity that a meteorite hailing from these regions would have as it entered our atmosphere. Most would burn away to dust without anything ponderable reaching ground level. Yet, we already know that Orgueil *was* in an eccentric orbit, but fortunately managed to avoid the usual dangers and entered our atmosphere at mild enough speeds to survive to the surface. Moreover, if we take into consideration the very small inclinations of the Themis family orbits (generally less than 3°), it seems entirely possible that Orgueil hailed from the Themids.

Yet, even if Orgueil did originate within this asteroid family, its association with comets is not totally severed. Themis asteroids are apparently quite comet-like in composition and, indeed, four of the family's members are already known to go through periods of cometary activity. It is widely thought that many more potential "main-belt comets" lurk within the family's boundaries. One of their number (176P/LINEAR) lies rather close to the 2:1 resonance. With a period of 5.72 years, it takes LINEAR 11.44 years to pass twice around its orbit, while Jupiter completes a single orbit of the Sun in 11.86 years. It is also interesting to note that, in common with Orgueil, LINEAR follows an orbit of unusually small inclination; a mere 14.12 *min*.

Let it be clear however, that I am not jumping to the conclusion that Orgueil originated as a chunk broken off main-belt comet LINEAR. But maybe there is an association, not just with LINEAR per se but with the Themis family at large. The bodies within this, or any, asteroid family were not all formed by a single impact. Over time, secondary collisions occur and larger family members get broken up into sub-families of their own. This is presumably how certain members of the Themis family can show cometary activity today; objects such as LINEAR were once parts of larger asteroids, buried so deeply beneath ancient surfaces that water ice remained stable during the billion-plus years since the collision that formed the original family. It is thought that the likes of LINEAR was broken away from a larger body through a secondary collision in *geologically* recent times and re-activated by a meteorite strike, gouging out a surface crater and exposing a layer of ice lurking not far beneath the surface of the fragment, in *historically* recent times. This is well demonstrated by the very first main-belt comet recognized as such; 133P/Elst-Pizarro, formerly noted as an asteroid of the Themis family but found to be sporting a tail 1996. Elst-Pizarro is now recognized as being a member of the Beagle family—a sub-family of asteroids within the larger Themis family and thought to have originated in a crash of asteroids around ten million years ago. The present activity of this comet, and presumably of LINEAR and the other Themis comets as well, is believed to have been triggered by exposure of sub-surface ice some time during the past millennium. Maybe even during the last century. It is suggested here that the sort of geologically recent asteroid collisions within the broader family that released small bodies carrying internal ice and capable of being activated as main-belt comets, may also have released boulder sized bodies, some of which dynamically evolved into orbits of the Orgueil type. One of these may even have been Orgueil itself!

A fascinating possibility raised here concerns the microfossil-like structures found in this and similar meteorites. If the controversial, but far from implausible, suggestion that these are exactly what they seem to be turns out to be correct, Themis-like and Ceres-like asteroids may not only be sites of life but may even have been the initial source of life on *Earth*; transported here by meteorites or even in the form of spore erupted into space from the interiors of these bodies during phases of cometary activity similar to that displayed by the likes of Elst-Pizarro and LINEAR. Now that is an interesting thought!

Far-Flung Mini Planets of the Outer Solar System

Prior to 1977, the outer planetary system was not known as the haunt of asteroids, except for the odd object 944 Hidalgo which we met earlier. But during that year, on November 1 to be exact, Charles Kowal, working with the Schmidt telescope at Palomar Mountain Observatory found an 18th. magnitude object that appeared to be an asteroid except for its unusually slow motion. It was announced in the *Circulars* of the International Astronomical Union as "Object Kowal" and simply described as a "slow moving object". There was nothing else to call it as the discovery of such a body was unprecedented. Orbital calculations revealed the object to be moving in a relatively low eccentricity orbit ($e = 0.38$), spending most of its time between the orbits of Saturn and Uranus. At perihelion (which last occurred on January 27, 1996 and will not happen again until 2046) the strange body lies at 8.5 AU from the Sun, just inside Saturn's orbit.

Today, Kowal's slow moving object is known as 2060 Chiron and also as 96P/Chiron, the former designation being asteroidal and the second is cometary. Chiron is one of the objects officially recognized as being *both* an asteroid *and* a comet! The cometary designation became necessary after Chiron brightened unexpectedly by 75 % in February 1988, developed a coma in April of the following year and sprouted a tail in 1993. None of this came as a complete surprise however, as Chiron had been suspected of being of cometary composition from the time of discovery. Chiron is a dark C-type object (sub-class B) and is estimated at around 146 miles (233 km) in diameter. Interestingly, its orbit is chaotic on a long time scale, making its distant future impossible to predict, although it seems that the inclination of its orbit will increase from its current 7° to about 9° and its period shrink a little to around 46 years by 7400 AD. If it does not encounter a major planet and become ejected from the Solar System sometime in the more remote future, it will probably end up as a short-period comet of Jupiter's family. Which of these fates await it cannot be predicted but given its considerable size, if it does finally become trapped in a short-period orbit it is likely to disrupt over time into an entire system of periodic comets and a debris complex larger than anything we experience today. If that is the ultimate destiny of Chiron, our distant descendents may experience spectacular short-period comets (maybe some that remain visible to unaided eyes right round their orbits!), massive meteor showers and a zodiacal light on steroids! But before we become too envious of these distant humans, we may ponder whether having all that cosmic debris floating around might have its drawbacks. Do we really want to live at a time when Tunguska-type impacts occur every decade or thereabouts? Far from being envious, we might start to wonder if humanity can survive Chiron! Hopefully, our descendents will have access to resources that we do not, should such a situation ever become reality.

At the time of its discovery, astronomers wondered whether Chiron was a lone freak or the forerunner of a new and previously unknown class of minor Solar System object (the correctness of even referring to it as an "asteroid" was even up for debate). Kowall apparently suspected that his object had siblings as he proposed that, since Chiron was named after one of the centaurs of Greek mythology, any other objects discovered with similar orbits should also be named after these mythological beings. Other similar objects have indeed been found and this practice followed. In fact, objects orbiting between Jupiter and Neptune are now collectively referred to as *centaurs*. As at 2010, some 183 have been listed and it is estimated that the total centaur population larger than 1 km in diameter is around 44,000. Chiron is anything but alone!

Another centaur that deserves special mention is 60558 Echeclus, also known as comet 174P/Echeclus. Discovered in 2000, this object maintained an asteroidal appearance until December 2005 when, like Chiron before it, it developed a cometary coma. Observations on December 30 revealed that a fragment had broken off the main body. Moreover, as astronomers continued to monitor developments, it became apparent that the center of activity was the fragment and not Echeclus itself. Apparently, a sublimating piece of ice and frozen gases had somehow become detached and was in the process of dissolving into a cloud of gas and dust. After a while, the fragment disappeared, the coma faded and the centaur regained its asteroidal appearance. At least, it regained this appearance until June 2011 when it erupted again into a new round of cometary activity. This time, no fragment was detected and the activity appeared squarely centered on the main body itself. The outburst was brief however, and already during July the spreading coma was becoming very diffuse and faint as it dispersed into space.

It will be interesting to see what happens to Echeclus in coming years. With a period of just over 35 years, it passes through its first perihelion since discovery on April 21, 2015 at a distance of 5.8 AU from the Sun. There will probably be further outbursts but it will be especially interesting to see whether activity becomes continuous for a significant period either side of perihelion. Suitably equipped amateur astronomers can play an important role in the monitoring of this peculiar object as, indeed, they did in the observations of its earlier outbursts.

Project 1: Keeping Watch on a Centaur

On April 22, 2015, the centaur Echeclus reaches perihelion for the first time since its discovery. Most of the time since discovery, this object has maintained a typical asteroidal appearance, but for a while in 2005 and again in 2011, it brightened dramatically and sprouted a coma. At brightest, it reached magnitude 14 or thereabouts—a brightness increase of some four magnitudes—and was observed visually through larger amateur telescopes. All of which raises the question: What will it do in 2015? Will it become continuously active around the time of perihelion?

Unfortunately, it will not be well placed at perihelion, but a careful monitoring with moderate-sized telescopes fitted with CCDs earlier in 2015 might catch the start of any activity that may occur. Following perihelion, Echeclus will emerge into the morning sky during May and it will be interesting to see how bright it then might be. Of course, there is no guarantee that it will be active and it must be said that if it reached the same level of activity at previous perihelia as it did during the outbursts of 2011 and, especially, 2005, there is a fair chance that it would have been discovered a lot earlier.

Continued

> **Project 1: (continued)**
> But only careful monitoring will sort out the behavior of this strange little world.
>   The orbital elements of this object are as follows. Updated elements can be accessed via the Internet closer to the time of the forthcoming perihelion passage.
>   Date of perihelion = 2015 April 22.5310
>   Perihelion distance = 5.817064 AU
>   Argument of perihelion = 162.9331°
>   Longitude of node = 173.3353°
>   Inclination = 4.3437°
>   Eccentricity = 0.455461°

Mini Worlds Afar: The Kuiper Belt and Beyond

Out past the realm of centaurs—beyond the orbit of Neptune—lies a broad band of myriad small objects generally known as the *Kuiper belt* or, sometimes, the *Edgeworth-Kuiper belt*, named after the astronomer(s) who first proposed its existence. Originally, this "belt" was hypothesized as a source for the numerous comets of short period trapped within the inner Solar System by the powerful gravitational pull of Jupiter. It was argued (correctly as it turned out) that such a relatively large number of comets could not have been captured from the field of long-period objects coming in from the still more remote Oort Cloud and it was felt that some closer reservoir more closely aligned to the plane of the planetary orbits, was required.

Needless to say, objects having the dimensions of the nucleus of an average short-period comet are extremely faint at Neptune's distance and beyond, but since the 1990s, search programs have turned up increasing numbers of bodies a lot bigger than comets but moving in the type of orbits expected for bodies within the Kuiper belt. Clearly, other things beside comets lurk out there in the dimness of the outer Solar System. Bodies having diameters of several hundreds of miles and larger have been found out there

In fact, one unusually large Kuiper-Belt object has been known since 1930. It is called Pluto!

Strange Little Pluto

The story of Clyde Tombaugh's discovery of Pluto in 1930 is well known; how apparent discrepancies in the motion of the outer planets led Percival Lowell and (independently) W. H. Pickering to calculate the position of an unknown "Planet X" in the way that similar studies of the motion of Uranus by J. Adams and U. Le Verrier the previous century successfully predicted the position of Neptune. These predictions led to the discovery of this planet and it was hoped that a similar search for Planet X would have similar results. When a faint slow-moving object was eventually discovered close to the predicted position of Planet X it was naturally supposed that the elusive world had been found. However, subsequent estimates of the mass of the new object—suitably named "Pluto" after the king of the Underworld—were unable to match that required to influence the orbits of the outer planets in the necessary way. The situation grew even worse as increasingly accurate estimates of its mass kept getting lower and lower—retreating ever further from what was required of Planet X. Eventually, reassessment of the masses of the outer planets themselves found that the supposed discrepancy disappeared and with it, the very *raison d'etre* for the existence of Planet X.

It seems that the discovery of Pluto had less to do with mathematical calculations of a hypothetical perturbing planet and more to do with the intensive search that these computations inspired. That the brightest of the Kuiper belt objects just happened to lie close to the computed position of the non-existent Planet X was one of those strange coincidences that happen from time to time in the history of science. Nothing is taken away from Tombaugh's discovery. Indeed even if he did not discover a ninth planet, what he did manage to achieve was the discovery of the first of an entire class of object that would not be fully appreciated for another 60-plus years. His was a discovery truly ahead of his time!

Nothing is taken away from Pluto, either, by its so-called "demotion" from the position of ninth planet. It is a strange and fascinating little world in its own right. If all goes well, the *New Horizons* space probe will reach it on July 14, 2015, so anything said about this distant world now will presumably (and hopefully!) be set for revision in the not-too-distant future. Nevertheless, a few facts about Pluto have, after decades of speculation, become relatively clear in recent years. (Incidentally, *New Horizons* carries some of the ashes of Pluto's discoverer Clyde Tombaugh, who passed away in 1997; a fitting if unusual memorial for this great astronomer).

The estimated density of this small world is 1.8–2.1 times that of water, suggesting a rock/ice composition estimated as 50–70 % rock and 30–50 % ice. Although uncertain for a long time and once thought to be at least equal to Mars (and even greater than Earth according to a minority opinion once voiced) its mass is now known to be less than 0.25 % that of our planet. With a diameter of some 1,440 miles (2,306 km) Pluto is nevertheless thought to be large enough to have generated sufficient internal heat to be a differentiated body. The model thought most likely at the moment is for a large rocky core of approximately 1,060 miles (1,700 km) in diameter overlain by a mantle of ice. If enough internal heat is still present, the boundary of core and mantle might be marked by an ocean of liquid water some 62–112 miles (100–180 km) thick (or *deep*). If that turns out to be correct, speculation about primitive Plutonian life might not be as absurd as generally supposed!

Over 98 % of the surface of Pluto consists of frozen nitrogen, with trace amounts of frozen methane and carbon monoxide. The overall color of the dwarf planet is red, albeit not as strongly red as Mars. Yet, a more detailed look at its surface reveals that differing regions display one of the greatest color contrasts to be found amongst known Solar System objects. In places, the surface of Pluto is charcoal black; in other places, dark orange and in yet other regions, white. Hopefully, we will know more about these color differences and the surface features with which they are associated after 2015. What is apparent however is that the color of Pluto's surface is not stable as the dwarf planet experienced a definite increase in redness between the years 2000 and 2012. Presumably, more will be learned about this also through space-based observations in the future.

This little world shares with the giant Uranus a total disregard of the ecliptic plane when it comes to axial orientation. Pluto reclines sideways with an enormous tilt of 120°. As a consequence of this, at the time of the Plutonian solstices, one quarter of its surface is in perpetual daylight (such as it is on Pluto!) and the other three quarters in perpetual night. Rotation around this tilted axis is slow, with 1 day on Pluto equivalent to 6.39 on Earth and, because the axial tilt is greater than 90°, rotation is considered to be retrograde. The cause of this extreme tilt is probably the same as that proposed for Uranus; a collision with a large body in Pluto's youth. There is further evidence that such a collision took place in the form of a system of moons thought to have coalesced from the debris thrown up by the event, but more about this in the following chapter.

The mention of "daylight" in the last paragraph must be taken in an extended sense. Moreover, unlike Earth where there is little difference in the apparent size or brightness of the Sun throughout the year, the orbit of Pluto is so eccentric that during its year (248.09 times longer than ours) the distant Sun varies in apparent size from about one minute of arc in diameter at perihelion to around 0.6 of an arcminute at aphelion and in brightness from about −19.5 at perihelion to a magnitude fainter (i.e. less than half as bright) at aphelion. Incidentally, the perihelic brightness is around 680 times fainter than experienced on Earth… hence my disparaging remark about Plutonian daylight!

An odd feature of Pluto is its slow evaporation away into space. It is, in a sense, a very weak comet and if much closer to the Sun, would sport a coma and even a tail. As it is, Pluto develops a transitory atmosphere of nitrogen, methane and carbon monoxide when close to perihelion, most of which re-freezes back onto the surface as the dwarf planet pulls once more away from the Sun. Some of the atmosphere does, however, escape into surrounding space in the manner of a comet's coma, and is lost to Pluto. The development of this atmosphere causes a drop in surface temperature for the same reason that alcohol rubbed onto our skin feels cold. This loss of heat through evaporation of surface ices means that, even as Pluto approaches perihelion, its surface cools. Infrared observations in 2006 gave a temperature 10 °C cooler than expected for a non-evaporating object at Pluto's distance.

Other "Plutos" and Their Siblings

Returning to Kuiper belt objects in general, several groups or families of these have been distinguished according to their orbital characteristics; principally how they relate dynamically to Neptune. Several resonances exist between the orbital periods of Kuiper belt objects and Neptune. Thus, those having periods of around 250 years complete two orbits for every three of Neptune and are said to be in a 2:3 resonance with that planet. Another group, with periods around 275 years has a 3:5 resonance with Neptune, completing three circuits for every five of Neptune. There are other resonances as well; for instance, objects with periods of 290 years are in a 4:7 resonance, those with 330 year periods in a 1:2 resonance and those around 410 years in a 2:5. Pluto, it should be noted, is in a 2:3 resonance with Neptune and therefore becomes the prototype of this family of Kuiper belt objects, not surprisingly known as *plutinos*. Members of the 1:2 resonance family are also given special names. They are *twotinos*. Then there are Kuiper belt denizens that are not in any resonance with Neptune at all. These are known as *cubewanos*.

Whilst speaking of cubewanos, it will be worth mentioning that the largest of their clan thus far discovered—136472 Makemake—appears to develop a transient atmosphere not unlike that observed on Pluto, albeit with a higher concentration of methane relative to nitrogen. This little world was discovered on March 31, 2005 by Michael Brown and evidence for a thin mantle of nitrogen and methane was soon forthcoming. The former gas escapes into space but at the temperature of Makemake (about 30–35 K), methane is not readily lost and therefore accumulates in the thin atmosphere. We will see later that this object—just 444–469 miles (710–750 km) or thereabouts in diameter is nevertheless large enough to join Pluto as another of that special class of minor objects: the dwarf planets (Fig. 3.6).

The Kuiper belt, properly so called, extends out to about 50 AU from the Sun. But that is far from being the limits of the Sun's family of small bodies, for overlapping the belt's furthest extremities and extending beyond them lies the realm of the so-called *scattered disk objects*; bodies that have been thrown outward from the planetary system by the gravitational perturbations of the giant planets.

156 Weird Worlds

Fig. 3.6 The dwarf planet Makemake as imaged by the *Hubble Space Telescope* (Credit: NASA/ESA)

Members of this population have perihelia greater than 30 AU and a broader scatter of orbital inclinations than the Kuiper belt inhabitants. The latter are more closely aligned with the plane of the major planets and form a torus-like system, with the scattered disk, in effect, fanning out at its furthest extremity. The largest body thus far discovered in the scattered disk is Eris, about which more will be said below. This large body is believed to have once been a Kuiper belt object, but, like the other denizens of the disk, was perturbed outward by planetary perturbations. Neptune is suspected as being the party responsible for this expulsion.

The Far-Out Oort Cloud

Not even the scattered disk marks the boundary of the Sun's family however. Beyond it, extending out to distances of at least 50,000 AU (about 1 light year!) is an approximately spherical halo of small bodies known as the *Oort cloud* or, more fairly, as the

Oort-Opik cloud. Actually, it would be even more appropriate to call it the Opik-Oort cloud as Ernst Opik postulated its existence before Jan Oort reached similar conclusions. Opik thought that it may have represented the remains of the original solar nebula. Oort found evidence of the cloud in his study of the orbits of a sample of comets with very long periods. Looking at the distances from whence these objects came, he found a noticeable clustering within a relatively narrow region so far removed from the Sun as to be barely gravitationally bound to the Solar System; distances of the order of 50,000 AU and greater. So weakly bound were these comets that any slight gravitational boost or retardation by one of the Sun's planets would have either sent them into interstellar space or else reduced the aphelia of their orbits to points well inside these distances. Therefore, Oort concluded, the majority of these objects must be entering the inner Solar System for the first time and the clustering of aphelion points at great distances presumably marks out the region of their origin. Moreover, as the scatter of their orbital inclinations was distributed right around the sphere, this remote "reservoir" of comets must be more or less spherical. Like Opik before him, Oort concluded that the planetary system is surrounded by an enormous spherical cloud of comets, most of which never come anywhere near the planets. The ones we see are simply those that have been deflected inward toward the Sun by the gravitational action of passing stars or molecular clouds. Others are directed in the opposite direction—away from the Sun –becoming part of the flotsam and jetsam of the Galaxy.

It is now thought, contra to Opik's hypothesis, that these objects did not form in situ within the Oort cloud, but were ejected there from smaller distances during the turbulent early days of the Sun's family. This is, indeed, similar to what Oort himself believed, although his hypothesis that they were ultimately fragments of a shattered planet has been replaced by a model which sees them as the building blocks of planets that never came together to form worlds in their own right. The Oort cloud is also thought to be more complex than initially understood. Oort's "dynamically new" comets (as they are generally termed) are now seen as coming from the *outer Oort cloud*. But this "outer" cloud is not stable over the life of the Solar System thanks to perturbations by giant molecular clouds and the Solar System's sporadic passages through these galactic structures. So tenuously held are the comets at these distances that they are likely to be stripped away by a crossing of a galactic spiral arm or the passage through a nebula. But as the old objects are stripped away, they are replaced by new denizens populating an *inner Oort cloud* and a disk-like structure known as the *Hills cloud*. Objects orbiting in these latter regions are normally unaffected by the perturbations of passing stars, but any disruption sufficiently major to strip away the outer cloud will also be major enough to cause the orbits of many objects within the inner cloud to expand sufficiently to take their place.

The process is one of outward diffusion of objects from the outer planetary system (probably near the orbits of Jupiter and Saturn originally) as they were hurled outward by the growing giant planets first into the inner cloud and then, through perturbations of giant molecular clouds, into the outer cloud and thence—via further perturbations—either into interstellar space or back again into the inner planetary system.

A possible association between the material composing Oort cloud objects and the ice of the moons of the gas giant planets is hinted at by the findings of the Cassini spacecraft which measured the D/H ratio (the ratio of ordinary hydrogen to "heavy hydrogen" or deuterium) in the water-vapor plumes spraying out of Saturn's moon Enceladus. It was found to have vastly more deuterium than Saturn itself or any of the other planets, but to have a D/H ratio very close to that of comets Halley, Hyakutake and Hale-Bopp, nearly as much as the 13-year periodic comet Tuttle and slightly more than the dynamically new comet 2002 T7 (LINEAR) and Ikeya-Zhang of 2002, a rather bright naked-eye object which turned out to be a return of the historical comet of 1661. These objects are all believed to have originated within the Oort cloud. The short-period comet 103P/Hartley which passed Earth rather closely in late 2010 and was imaged close-up by NASA's *EPOXI* spacecraft and observed by the infrared *Herschel Space Observatory* in November of that year was found, by the latter observatory, to have a D/H ratio just half that of Halley and Enceladus. As mentioned earlier, the D/H ratio of this object turned out to be exactly that of Earth's water and of carbonaceous meteorites. It is noteworthy that Hartley (as a member of Jupiter's comet family) almost certainly came from the Kuiper belt initially. In terms of the D/H ration therefore, Saturn's Enceladus seems to be more closely related to the Oort cloud than to the Kuiper belt!

At the distances of the outer cloud, its denizens are far and away too faint to be seen by any instrument on Earth. Only by observing the very long-period comets after leaving the cloud can its existence be ascertained. Typical comets are only a few miles across (this refers only to their nuclei of course as the much larger coma and tail structures do not exist at the distances of the Oort cloud) but larger bodies might well lurk out there, totally unobservable from Earth. Pluto, for example, would be as faint as magnitude 46 or thereabouts at the distance of the outer cloud!

The largest members of the innermost regions of the inner cloud, just a few 100 AU out from the Sun, may be within our ken however, and it has been suggested that some objects already listed in our catalogues might justifiably be classified as bona fide members of the inner Oort cloud. These objects lie beyond the Kuiper belt and are simply classified as *Trans Neptunian Objects* (TNOs); the general term under which all objects outside of Neptune's orbit are conveniently lumped together. However their distance is such that they may justifiably be assigned to the inner cloud. The list of candidates consists of Sedna, 2000CR105, 2006SQ372 and 2008KV42. These are all rather large bodies. As we shall see shortly, Sedna is high on the list of probable dwarf planets. The discovery of these bodies implies that large ones exist in the inner cloud and, insofar as the outer feeds on the inner over the life of the Solar System, this further implies that bodies of dwarf planet scale exist in the outer cloud as well. The possibility of full scale planets within the Oort cloud is not impossible, although the failure to find any amongst the nearer classes of TNO may make their presence less likely.

Damocles… and a Bullet Asteroid in 2024

Whether large or small, the denizens of the Oort cloud have traditionally been thought of as icy bodies. But maybe that is not always the case. There is some evidence for the existence of rocky objects there as well. Thus, in 1996 an object designated 1996

PW was discovered in an orbit typical of a comet coming in from the Oort cloud. But the problem was, 1996 PW remained completely asteroidal in appearance throughout, showing no signs of developing cometary activity. Although the possibility remains that this object was a defunct comet that had been ejected to Oort cloud distances by planetary perturbations, it is at least equally possible that it was a rocky asteroid which never possessed cometary potential. Indeed, we now know of an entire class of asteroids which, whilst not moving in orbits as extreme as that of 1996 PW, nevertheless orbit the Sun along paths that are more comet-like than asteroid-like, even to the degree of being retrograde in some instances (i.e. having orbital inclinations greater than 90° and therefore effectively moving opposite to the planets). These bodies have been given the rather ominous sounding title of *Damocloids*, after their prototype 5335 Damocles, discovered in 1991 by Rob McNaught of Siding Spring Observatory in Australia. This object moves in an elongated elliptical orbit having aphelion at 22.1 AU and perihelion at 1.58, with an orbital inclination of 61.95°. This is very similar to a comet orbit, yet Damocles gave not the slightest hint of activity. Some 77 members of this class of strange objects have been recognized as at early 2012. Four of them have been shown to be very dark objects with a slight reddish coloration. They may be inert bodies captured from the Oort cloud or they may be worn out comets—or maybe a mixture of both. The suggestion that some may once have been active comets is apparently supported by the discovery of several objects that were initially thought to be Damocloids but subsequently reclassified as comets after they began to exhibit a low level of activity. These are cited as comets that have not quite gone dormant, although it is also possible that they are rock/ice objects that have never been quite icy enough to exhibit fully fledged cometary activity, yet too icy to remain totally *in*active. If the Oort cloud contains both icy and rocky objects, it is unlikely that there will be a clear and absolute line of demarcation between them. It is far more probable that one class will merge into the other with a borderland of transition objects capable of varying degrees of activity on approach to the Sun.

Whilst on the subject of Damocloids, mention must surely be made of the small asteroid 2009 HC82. Although its aphelion distance is just 4.67 AU, it is classified as a member of this peculiar asteroid class due to its retrograde orbit (inclination 154.52°). It has a rather small perihelion distance of 0.49 AU, but its chief claim to fame is its high velocity relative to Earth. Indeed, relative to our planet, it has the potential of being the fastest known object venturing closer than 0.5 AU. On November 11, 2024, this body will pass Earth at a distance of 0.485 AU. This is not *very* close, but because the asteroid and our planet will be moving in opposite directions, 2009 HC82 will pass us by at the record relative velocity of 176,813 miles (282,900 km) per hour!

Parabolic Planets?

Worlds the size of Earth or Mars may not lurk in the outer Oort cloud of our Solar System. Even if they do, they are surely few in number and it is unlikely that there has ever been—or ever will be—a full sized planet deflected into the inner planetary system along the near-parabolic trajectory of a dynamically new comet. This is not something that I regret, but the chances that it does occasionally happen somewhere in the vast Universe are, I would opine, pretty high. Solar systems harboring wandering Jupiters undoubtedly hurl terrestrial planets out into interstellar space but every so often some of these planets surely stop short of complete ejection and manage to end up in distant orbits around their suns—even as far out as the Oort cloud is in relation to our own Sun. From there, very occasionally, we might imagine one suffering the fate of a dynamically new comet and being deflected into the central reaches of its solar system.

What would it be like to be an immortal observer on the surface of one of these worlds watching the changes taking place during the hundreds of thousands or millions of years taken by such a world to drift from Oort cloud remoteness to terrestrial planet nearness of its sun? (Of course, unless the Universe houses a form of life totally beyond our comprehension, no real conscious organism could exist on worlds such as these).

For an Oort cloud object orbiting a sunlike star, its own sun would be just another bright star in the sky. If a blue super-giant was located relatively nearby, the object's sun would not even be the brightest star visible. But as the planet drifted inward, its sun would appear to brighten; ever so slowly at first, but as time stretched into aeons the pace of brightening would eventually start to quicken. Eventually outshining even the blue supergiants, this whitish brilliant in the sky starts to penetrate the deep gloom of an airless landscape of frozen deserts and mountains of ice. As our wandering planet approaches its star to the equivalent of the mean distance of Pluto from our own Sun, its star (assuming dimensions and intrinsic brightness similar to the Sun's) shines at a brilliance of around 270 times that of Earth's full Moon, albeit from a tiny disk just 45 arcseconds across. Soon something new begins to happen; something that has probably never occurred on this planet before. Frozen gases start to sublimate as the distant sun sheds enough warmth to wake up the dormant ice. Nitrogen, methane and other very volatile substances begin to turn into gas and a thin atmosphere slowly forms.

At the equivalent distance of Saturn, the planet's sun has increased its apparent size to 3 arcminutes and shines with the brilliance of over 4,000 full Moons. The atmosphere has become appreciable and we may imagine that the ultraviolet radiation from the sun has become sufficiently strong to break down and recombine molecules in the gaseous envelope, even producing something similar to the haze of organic particles that shrouds Saturn's giant moon Titan in our own planetary system. Maybe some form of "weather" is now occurring. Could convection in the atmosphere be forming clouds of methane droplets?

Nearer still and more ices sublimate. Atmospheric pressure builds higher and water ice melts. Ice deserts turn into oceans. Rain falls. Methane, water, nitrogen—what a mixture! What a pity that the planet now swings past perihelion and the process goes into reverse. Oceans that have just melted freeze again. Eventually, even the atmosphere freezes out as the sun slowly returns to being just another star. Would it not have been interesting had the planet been captured into a circular orbit about 1 AU from its sun. What might have happened then over the next several millions of years?

Weird Comets That Dive-Bomb the Sun

Before leaving the subject of Oort cloud denizens, let's take a little time to speak about one very odd group of bodies originating there. Although not strictly speaking "worlds", they are in their way, "weird" enough to deserve mention and really take us from one extreme of Solar-System distance represented by the Oort cloud to the absolute opposite.

The objects to which I am referring are the comets collectively known as *sungrazers*. The name tells us the nature of their common characteristic; very, *very* close approaches to the Sun. Since 1996, the SOHO space-based solar observatory has discovered well over 2,000 of these objects—all of them very small and all of them boiling away completely near perihelion. Add to that number several found by other space-based solar observatories and several more bright objects seen from the ground in recent centuries (some of these being the most brilliant and amongst the most spectacular comets on record) and we can see how numerically important these comets are. In fact, it would not be exaggerating to say that since the advent of SOHO, the majority of recorded comets have been sungrazers! And the most recent one discovered and observed from the ground—an added bonus during the festive season of 2011—taught us to be cautious about making definitive predictions, even those concerning minor objects in our own cosmic backyard! More about this object shortly.

Prior to the launch of orbiting solar observatories, a cluster of three well established sungrazers appeared in the 1880s and a second equal cluster in the 1960s. There was also an isolated very bright one in 1843 and another isolated one (this time only faint and briefly observed) in 1945. A comet seen only during a total solar eclipse on May 17, 1882 may have been another and there were more suspected members in the 1600s and earlier, together with a very likely suspect in early 1702. The three most brilliant were clearly visible with the naked eye in full daylight near the limb of the Sun. These were the comets of 1843, the "September comet" of 1882 and Ikeya-Seki of 1965, the brightest comet seen during the last century. This latter object was the only one actually tracked right through perihelion passage and became briefly brighter than the full Moon, almost certainly brighter than any other comet actually under observation, although those of 1843 and 1882 may have equaled it had they also been observed through perihelion. Yet, at the time of greatest brilliancy it was so close to the Sun that only the astronomers at Tokyo Observatory's coronagraph station managed to observe it. The September Comet of 1882 was, however, the largest and intrinsically brightest of modern sungrazers. It is interesting to speculate that, had its brightness increased at the average rate all the way to perihelion, it would have brightened to within seven magnitudes of the brilliance of the Sun itself. But because of its small diameter as measured just before disappearing against the disk of the Sun several hours earlier, its brightness intensity would actually have reached some 390 times greater than that of the Sun's disk! As it was not seen near the Sun at perihelion, we can be sure that it actually came nowhere near this brightness or intensity. It must have reached an upper limit well shy of these values, probably similar to that later observed for Ikeya-Seki.

Defining "sungrazer" as strictly applying to comets having perihelia of less than 0.01 AU (those between 0.01 and 0.1 are now often referred to as *sunskirters*), we are immediately struck by the fact that, of the 2000+ known today, all but three followed very similar orbits, strongly hinting at a common origin. Of the three that did not, only one was observed from the ground. This was the historic Great Comet of 1680. The other two lone sungrazers were very faint objects seen only in SOHO data, one in 2001 and the other in 2007. Other than perihelion distance, their orbits bear little resemblance either to the sungrazer majority, or to the 1680 comet or to one another.

The sungrazing "group" was first investigated by Heinrich Kreutz (1845–1907) in the late nineteenth century, following the comet cluster of the 1880s. His conclusion that they are all fragments of a single disrupted comet has not been seriously challenged, although it has evolved over the decades, in part due to the discovery of further members and in part because of more powerful computing techniques.

The most recent and comprehensive study of these strange objects is a paper by Z. Sekanina and P. Chodas published in *The Astrophysical Journal* for July 1, 2007. Like earlier studies, this model sees the group as derived from progressive splitting of several generations of comets leading back to a single progenitor. Unlike other studies however, it takes into account not just tidal splitting close to perihelion, but non-tidal splitting at any point in the comets' orbit.

Briefly summarized, the Sekanina-Chodas model is as follows. The clusters of nineteenth and twentieth century sungrazers (of which the 1843 and 1945 comets appeared to be forerunners of the two respective clusters) and probably earlier ones as well, were caused by the non-tidal splitting (far from perihelion) of fragments whose origin lay in the tidal (perihelic) splitting of a previous generation of sungrazers. Most of the nineteenth century cluster, though paradoxically, not its brightest member—the September Comet of 1882, which had a *slightly* different orbit—originated in the tidal disruption of a comet that appeared early in the twelfth century. The authors identify this comet as the daylight object of 1106. The September Comet of 1882 and Ikeya-Seki of 1965, each of which shared very similar orbits, resulted from a very similar the disruption of another comet which also must have reached perihelion in the early twelfth century, but was apparently badly placed and passed unrecorded. The 1106 comet, at a previous apparition, may have been identical with the one seen in February 423 or the one in February 467. Sekanina and Chodas favor the latter and opine that this comet and the progenitor of the 1882/1965 pair probably split from a Grand Progenitor in a non-tidal disruption far from the Sun whilst on its way to the fifth century perihelion. This object, the parent of the entire sungrazing group, may have been at its previous perihelion passage, the one recorded very briefly by the Chinese in the year 214 BC. Unfortunately, we know next to nothing about this object, except that it must have been at least reasonably bright, that it had a tail (the record itself simply calls it a "star", but a commentary refers to it as a "broom star" indicating the presence of a curving tail) and that it appeared in the western (presumably evening) sky. Alas, the time of year is not given, but if it was during the first half its evening appearance is consistent with a "Kreutz" sungrazer. If it was in the latter part, a Kreutz association is precluded.

One exciting conclusion reached by these authors in their 2007 paper was that prospects seem good for the arrival of another cluster of bright sungrazers, with the earliest member(s) possibly only years away. This prediction now appears to have been fulfilled!

The Latest Example of a Major Sungrazer

The most recent member of the Kreutz comet family visible from the ground was C/2011 W3 (Lovejoy), discovered by Australian amateur astronomer Terry Lovejoy in November 27, 2011 patrol images of the southern sky. Observations through early December showed the new object to be a sungrazer; the first discovered from the ground since 1970. Perihelion was computed to occur very early on December 16 at which time the comet was due to skim over the Sun's surface just 80,000 miles (130,000 km) or thereabouts above the photosphere at a velocity exceeding one million miles per hour.

Now, by our normal standards of distance, 80,000 miles seems quite far, but when this measures the distance from the surface of the Sun, it is an entirely different matter. The equivalent proportionate distance above Earth's surface is around 800 miles. Ikeya-Seki by contrast "only" approached to about 300,000 miles (470,000 km in rounded terms) from the Sun's photosphere; proportionate to some 23,000 miles from the surface of Earth. At Lovejoy's distance from the photosphere, the tops of solar prominences, had any been present close by, would have towered above it, the taller ones exceeding the comet's altitude by several hundreds of miles!

Not surprisingly in view of the comet's faint intrinsic magnitude at discovery and extreme orbit, nobody expected it to survive perihelion.

Come December 15 and comet enthusiasts around the world were glued to their computer screens as images from SOHO were displayed showing the (supposedly) doomed comet plunging toward the Sun. It brightened rapidly in the early SOHO images but, just hours before perihelion, it started to fade; a common feature of the mini-comets observed in SOHO data and one which appeared to confirm Lovejoy's impending demise. Further apparent confirmation of the comet's disintegration came soon after the faded head vanished behind the solar occulting disk of SOHO's coronagraph. The bright tail clearly became detached from the (presumably defunct) head and began to drift slowly away from the Sun as it faded and dispersed.

So goodbye comet Lovejoy? Well, not exactly as it turned out!

Early on December 16, as computer-screen-bound comet observers watched the fading tail drift away from one side of the occulting disk, a totally unexpected star-like object rocketed out from behind the other side and moved quickly away in a direction opposite to that of the tail. This was the emaciated head of the comet, devoid of any trace of tail and initially as faint as a star of fifth magnitude. Somehow, in defiance of all the sound arguments of the experts, a solid remnant of the nucleus had survived its fiery plunge through the heart of the Solar System. Moreover, it did not remain faint for long. While still early on the 16th the star-like object went gangbusters, surging in brightness until it at least rivaled Jupiter. It was then more brilliant than it had been at any time prior to perihelion. A new tail also began to develop by the end of the day and on the 17th. became a prominent feature in SOHO images. Several telescopic observations—both photographic and visual—were made in broad daylight shortly thereafter and by December 21, the tail had become a spectacular sight rising out of the dawning sky for southern hemisphere observers. Its appearance resembled (and on at least one occasion was briefly mistaken for!) the beam of a searchlight. In the dark pre-dawn of Christmas morning, over 20° of tail could readily be seen with the naked eye and by the first days of January 2012—although by then appreciably fainter—naked-eye observers blessed with clear and dark skies were tracing it for over 40°. Not since 1965 had a sungrazing comet provided such a spectacle! (Fig. 3.7)

Before leaving the "Great Christmas Comet of 2011" we might again note that its appearance seems to have at least begun to fulfill the 2007 prediction by Sekanina and Chodas that the first member of a new cluster of sungrazers may only be years away. This is an encouraging sign for comet hunters.

Fig. 3.7 Sungrazing comet C/2011 W3 (Lovejoy) from International Space Station, 21 December 2011 (Credit: NASA/Dan Burbank)

Project 2: Hunting Down the Sungrazers

In these times of professional sky patrols, it is heartening to remember that the "Great Christmas Comet of 2011" was discovered, and named in honor of, an amateur astronomer. If other bright sungrazing comets do arrive in the years ahead, they probably offer a better chance than most of being first picked up by amateur astronomers. Being fragments of a larger object, the Kreutz members tend to be small and, because they have already been baked near the Sun on previous returns, the

Continued

Project 2: (continued)

surfaces of their nuclei are, in many instances at least, probably depleted in volatile material preventing them from fully "switching on" activity until they have already approached the Sun quite closely. In other words, they are likely to remain very faint until they have ventured well within Earth's orbit. Most sungrazers discovered from the ground have made a sudden appearance close to, and more often following, perihelion. Nowadays, thanks to space-based solar observatories, there is little chance of ground based astronomers—be they amateur or professional—finding a sungrazer very close to perihelion except, of course, on the web pages of these observatories. But in a paradoxical sort of way, the faint nature of these comets as they approach perihelion works decidedly in the amateur's favor. The professional search programs concentrate on regions of the sky at relatively large elongations and most sungrazing comets passing through these regions will still be very faint and in all probability undetectable even by the instruments employed in these searches. It is only when they are at much smaller elongations, just a few weeks before perihelion, that they will rapidly brighten (as Comet Lovejoy did)—and it is at this point that the amateur astronomer has a real chance of finding them.

Below is a list of positions for the first and fifteenth day of each month for a hypothetical sungrazer coming to perihelion on the first day of the following month. The positions are equinox 2,000 and for zero hours Universal Time on the days concerned. In addition, the elongation (angular distance from the Sun) is given, plus a magnitude estimate based on the assumption that the comet has an absolute magnitude (H10) of 10. The "absolute magnitude" is the brightness that a comet would have if located at 1 AU from both the Sun and the observer. The choice of 10 is purely arbitrary. It is simply a nice round figure to act as a benchmark. Prior to their perihelia, Comet Lovejoy was about five magnitudes fainter and Ikeya-Seki four brighter than this value. There is a small

Continued

Project 2: (continued)

divergence in the orbits of Kreutz sungrazers, but most are closer to that of Comet Pereyra than to either Ikeya-Seki or Lovejoy. If another cluster really is on its way, it is likely that the orbits of its members will more or less resemble that of Pereyra and for that reason the positions below have been computed using the orbital elements of that comet.

These positions may be used to plot the area that members of the sungrazing comet group will traverse at different times of the year. The given dates should enable an area of sky to be marked out for searching. An Internet search will turn up information for more accurate plots if so desired. Any plot should, however, be seen as approximate and searches need to be conducted several degrees either side of the predicted path in order to cover the possible range of tracks that the comets may pursue. Photographic searches stand a better chance of success as they are capable of reaching to fainter magnitudes, but wide-field telescopes and large securely mounted binoculars are certainly not to be despised.

It will be noticed that southern searchers are more favored (though not exclusively so), that the search areas shift from the evening sky early in the year to the morning during the later months and that the mid-year period is a very unfavorable time for finding and observing members of the Kreutz group

Perihelion date	Position (1st and 15th. day of prev. month)					El	Mag. (H10=10)
Jan. 1	Dec. 1	11 h	19 m	−52°	50 m	74.2°	9.6
	Dec. 15	15 h	44 m	−66°	51 m	46.7°	7.4
Feb. 1	Jan. 1	3 h	9.9 m	−77°	56 m	78.3°	9.2
	Jan. 15	22 h	56.6	−50°	22 m	47.2°	7.7
Mar. 1	Feb. 1	1 h	39.9 m	−35°	45 m	64.5°	9.6
	Feb. 15	0 h	32.0 m	−22°	50 m	40°	7.8

Continued

Perihelion date	Position (1st and 15th. day of prev. month)					El	Mag. (H10=10)
Apr. 1	Mar. 1	2 h	36.8 m	−17°	2 m	56.7°	10.6
	Mar. 15	2 h	8.3 m	−9°	12 m	37.8°	8.9
May 1	Apr. 1	3 h	26.4 m	−5°	9 m	42.5°	11.0
	Apr. 15	3 h	18.7 m	+1°	5 m	28°	9.1
June 1	May 1	4 h	24.4 m	+1°	21 m	30.9°	11.5
	May 15	4 h	32.4 m	+6°	28 m	20.2°	9.6
July 1	June 1	5 h	23.8 m	+5°		20.7°	11.6
	June 15	5 h	45.0 m	9°	12 m	14.4°	9.5
Aug. 1	July 1	6 h	23.7 m	+5°	17 m	18.2°	11.7
	July 15	6 h	56.6 m	+7°	59 m	16.7°	9.7
Sept. 1	Aug. 1	7 h	24.1 m	+2°	58 m	25.0°	11.6
	Aug. 15	8 h	9 m	+3°	49 m	24.1°	9.5
Oct. 1	Sept. 1	8 h	23.8 m	−2°	13 m	35.6°	11.3
	Sept. 15	9 h	24.6 m	−3°	47 m	32.2°	9.0
Nov. 1	Oct. 1	9 h	14.8 m	−11°	6 m	48.5°	11.0
	Oct. 15	10 h	33.4 m	−16°	9 m	41.2°	8.7
Dec. 1	Nov. 1	10 h	6.2 m	−26°	9 m	61°	10.2
	Nov. 15	12 h	15.3 m	−36°	33 m	44.3°	7.8

Strange (Not So) Little World-Lets: The Dwarf Planets

Having ventured to the outer fringes of the Solar System and back again into hellish regions closer to the Sun's surface than the tops of solar prominences, it is time to take a closer look at a class of object met along the way but, until now, simply lumped together with other minor bodies.

The discovery of this class of relatively large "minor objects"—what we might call "Pluto-like bodies"—at the fringes of the Solar System raised questions as to how astronomers might classify them. Relatively small objects beyond the orbit of Neptune presented no fundamental classification problem, as they were simply dealt with in the manner of minor planets (asteroids), even though they were likely to be icy rather than rocky in composition. But the larger bodies look very much like little planets. They are ellipsoidal in shape and seemed uncomfortably close in nature to Pluto itself. If Pluto is classified as a planet, the objects that first started turning up in the 1990s should be planets as well. Maybe a single discovery or just one or two would have seen these bodies added to the list of the Sun's major companions. Ten, eleven or twelve planets in the Solar System looks alright—but if the number climbs into the hundreds or even thousands, there seems to be a need to differentiate these small objects from the larger ones which from the core of the planetary system. But if they are classified as something other than planets, where is the justification for retaining Pluto as the ninth planet, as astronomy texts have been teaching us ever since the 1930s?

To make a long and at times quite heated story short, the upshot was the recognition of a new class of *dwarf planets*. These objects are "planets" in a narrower sense than the wider "minor planet" term used for asteroids and similar small bodies. These latter are "planets" only in the sense that they orbit the Sun. True dwarf planets also possess the planetary property of being in hydrostatic equilibrium, i.e. being massive enough for their shape to be controlled by gravitational rather than mechanical forces. They have, therefore, settled down into ellipsoidal shapes rather than having the irregular forms of typical asteroids; shapes more often determined by past collisions than by any equilibrium process. To that degree, these are true planets, however they are differentiated from the "major" representatives of this class by being insufficiently massive to have cleared their neighboring regions of other co-orbiting objects. The lower limit for a rocky object to achieve the required hydrostatic equilibrium is about 375 miles (600 km) and for an object that is mainly water ice, about 200 miles or 320 km. The upper limit (which is also the lower limit for major planets) is not well defined and even something as large as Mercury that had failed to clear its orbit could justifiably be called a dwarf planet. Mercury itself seems to have done its clearing job sufficiently well and is in no danger of being shifted into the dwarf planet class—which is just as well, considering the hullabaloo that followed Pluto's shift of status (it is not often that a decision by the International Astronomical Union is met with placard-waving protests, but its decision on Pluto managed it!).

As at early 2012 when these words are being written, the number of *well established* dwarf planets stands at a modest five; Ceres, Pluto, Haumea, Makemake and Eris. Ceres is the smallest and the only one not classifiable as a TNO. At present, the honor of being the largest known dwarf planet is being fought out between Pluto and Eris. Estimates of the latter's diameter range from slightly smaller to slightly larger than that of Pluto and it is probably best to think of them as essentially the same size until more accurate observations can be made. It does seem, nevertheless, that Eris is about 27 % more massive than Pluto and is therefore a denser object, probably having a higher rocky content than its fellow dwarf planet.

The true number of these objects must far exceed a mere five however. After making assumptions regarding average reflectivity of trans-Neptunian objects and so forth, it is estimated that an object large enough to cross the threshold from mere minor-planethood to true dwarf-planethood should have an absolute magnitude of +1. Therefore, it is assumed that anything equaling or brighter than this will be a dwarf planet. On the basis of this assumption, the number of suspected dwarf planets listed in Mike Brown's catalog of Kuiper belt and other Trans Neptunian objects, as at April 2011, reached 390. Brown estimates that there are likely 200 dwarf planets in the Kuiper belt strictly so called and possibly in the order of 2,000 in the wider population of TNOs. Beside the five listed above as confirmed members of the class, four further objects that should very probably be included are Orcus, Quaoar, 2007 OR10 and Sedna. The last of these is interesting in that it orbits far beyond the Kuiper belt and has an *average* distance from the Sun greater than any other TNOs thus far discovered, being almost 519 AU from the Sun and taking approximately 11,400 years to make a single circuit of its orbit. (At present however, Eris is actually more remote as it is close to the aphelion of its orbit whereas Sedna is approaching perihelion.) Sedna probably resembles an over-sized comet nucleus and is estimated to have a diameter of about 870 miles (about 1,400 km). If these estimates are correct, it lies well within the range of the dwarf planet category.

With distant Sedna, we leave the minor members of the Solar System. No longer dismissed as "clods of dirt", "bagfuls of nothing" or "vermin of the skies" these are fascinating and, in their own way, strange little objects. We have seen that some are true little planets while others better resemble flying mountains. Some develop temporary atmospheres and same may even harbor underground oceans where (just maybe) simple forms of life could have gained a foothold. Unlike the major planets of the Sun's family, the minor bodies range far and wide. Some orbit in perpetual deep freeze at distances where the Sun is nothing more than the brightest star in the sky, while others zoom in so close as to briefly venture down amongst the solar prominences. What other odd-ball minor objects might be found in the future, only time will tell. But for now, let us move on to another class of Solar System objects which, in common with the subjects of the present chapter, have (with one major exception) largely been overlooked until the advent of space exploration. We refer to the class of moons or satellites, some of which have turned out to be, on close inspection, amongst the most interesting of objects with their own individual characteristics and their own unique "weirdness".

4. Moons Galore!

Most of the planets and many of the asteroids and icy denizens of the outer Solar System are far from being lonely bodies floating through the vastness of space. Moons or satellites are a frequent occurrence amongst the members of the Sun's family and range in size from the space rocks found orbiting some asteroids to orbs the size of small planets. Two moons (Jupiter's Ganymede and Saturn's Titan) are larger than Mercury, and another Jovian moon (Callisto) is only a little smaller than that planet. If these bodies were in solar orbit instead of tied to a gas giant, they would certainly be included as fully fledged planets—not even dwarf planets, but the real thing! In fact, Ganymede is rated as the ninth and Titan as the tenth largest known object in the Solar System, exceeded only by the Sun and the planets from Venus to Neptune (comet heads or comas are not included as these are not self gravitating bodies).

The two planets nearest the Sun are the exceptions to the satellite rule. No natural satellite has been discovered orbiting either Mercury or Venus and if any exist, they must be very tiny. Although it is not impossible that one or more diminutive asteroids or large meteoroids have been captured by these planets, at the present time there is no evidence to this effect and each of these worlds seems to be completely moonless. Certainly, the early reports of a supposed satellite of Venus must be relegated to the limbo of mistaken observations.

The Moon

For denizens of planet Earth, "moon" primarily means *the* Moon; the single permanent natural satellite of our world. After all, it appears to us as the second brightest object in the sky and far outshines all other nocturnal objects, periodically brightening our nights (not always welcomed by those astronomers whose principal field of interest concerns faint objects easily drowned out by strong moonlight!) and is the chief regulator of our planet's tides. More recently, good arguments have been proffered to show that the Moon's stabilizing effect on Earth's axial tilt and even its influence on the velocity of our planet's rotation may play a non-inconsequential role in the very habitability of our world, insofar as highly complex organisms are concerned at least.

Although moons are common, the Earth-Moon system is in certain important ways an exception to the rule. Leaving aside for the present the dwarf planet Pluto and its oversized satellite Charon, the Earth is the only known major planet to sport a moon that is large when compared to the primary itself. Although there are four larger moons within the Solar System, these are found in orbit around the giant planets where, by comparison with their primaries, they appear relatively insignificant. Our satellite however, has a diameter of nearly 0.3 times that of Earth itself. For a moon/primary system, that is a most unusual ratio of sizes. Amongst the small rocky planets, it is doubly unusual; as we have seen, neither Mercury nor Venus has any moon at all and Mars is only accompanied by two flying mountains that hardly compare with our huge moon. Indeed, some have even suggested that what we call the Earth/Moon system is really nothing of the kind! Instead of being numbered amongst the primary/satellite systems, they suggest that our planet and its so-called moon would better be described as a double planet; the only one of its kind in the Solar System. We can see where these folk are coming from, but their suggestion is not entirely accurate. A true double or binary planet would see two worlds in orbit around a mutual center of gravity located outside of the body of each planet. Currently, we do not know any full-sized binary planets. Certainly, this situation does not occur with the Earth and Moon. The center of gravity remains within the confines of Earth, such that the Moon is still effectively a satellite of Earth, despite its size. It is not quite massive enough to be a true companion planet. At least, that is the official line! However, the situation will theoretically change billions of years from now. Eventually the Moon will recede so far from Earth that the center of gravity of the system drifts outward to more than one Earth radius. So much will happen to the Sun and even the Galaxy in the meantime however that this is looking just a little too far into the future!

Nevertheless, the (present-day) location of the Earth/Moon system's center of gravity, has not daunted all supporters of the Moon-as-planet notion. Back in 1975, science writer Isaac Asimov drew attention to the fact that the Moon's orbit around the Sun is in tandem with that of Earth such that for a hypothetical observer looking down on the ecliptic, the Moon would never actually close its orbit and loop back on itself. In essence, it would appear to orbit the Sun in its own right! For the present purposes however, we will remain conservative, uphold the status quo, and class the Moon as a true satellite and not as a companion planet!

The Moon is a rocky body having a diameter of 2,171 miles (3,474 km) a density slightly in excess of 3.3 times that of water (which, by the way, makes it the second densest known satellite—second only to Jupiter's Io) and orbits Earth at an average distance of 240,249 miles (384,399 km) or 0.00257 times Earth's mean distance from the Sun, i.e. 0.00257 AU. Thanks to the tidal interaction between Earth and Moon over long periods of time, the latter's period of rotation on its axis is about equal to the time it takes to make one revolution of the Earth, so one hemisphere is perpetually turned toward and the other perpetually away from us. Therefore, although one face of our satellite has been known since time immemorial, it was not until the advent of space flight that the "other side of the Moon" became accessible to humanity. Not surprisingly, the "other side" did not turn out to be very different from the Moon's familiar face, at least in broad terms. The occasionally aired romantic notions of "the other side of the Moon" being verdant and welcoming were not true, but of course these had long been consigned to the realm of fantasy by astronomers. On the other hand, the far side differs in having far fewer of the dark plains known as maria ("seas") than the familiar near side (Fig. 4.1).

The Moon possesses a small solid inner core; probably about 300 miles (480 km) in diameter. The innermost core is surrounded by an outer one, bringing the entire core diameter to around 412 miles (660 km) or barely 20 % of the size of the entire body. This is in contrast with dimensions of around 50 % for most of the rocky planets and strongly contrasts with the 85 % or thereabouts for Mercury. Planetary scientists are still not sure of the lunar core's exact composition, but it is thought most likely to consist of iron

Fig. 4.1 Farside of the Moon (Credit: NASA)

alloyed with small quantities of sulfur and nickel. Observations of the Moon's rotation indicate that the outer core is in a fluid state.

The core is, in turn, surrounded by a partially molten boundary layer extending outward to just over 1,200 miles or about 2,000 km. This is thought to have formed through the fractional crystallization of an ocean of magma soon after the Moon formed. During this early period, heavy materials such as olivine, clinopyroxene and orthopyroxene sank toward the core while less dense materials rose to float on the surface of the magma ocean. By the time that around three quarters of the magma ocean had crystallized, these lighter (floating) materials had also largely solidified, forming a crust on the surface of the remaining magma. The very last liquids to crystallize would have found themselves squeezed between the mantle (the solidified magma ocean) and crust. The remaining liquid is believed to have been a mixture of incompatible and heat-producing elements which erupted through the crust in many places to produce the very iron-rich, basaltic, flow lavas which were sampled by the *Apollo* astronauts and robotic probes.

The dark plains or *maria*—so called because early astronomers mistook them for bodies of water—are, in a sense, "seas" after all albeit solidified seas of this basaltic lava, not of water. As already mentioned, most of these lava plains are on the visible near side of the Moon, presumably because most of the heat-producing elements concentrated under the crust on the hemisphere facing Earth. This is also indicated by the results of the gamma-ray spectrometer on the *Lunar Prospector*. The lava itself is not too unlike its terrestrial counterpart, albeit rather richer in iron and totally void of any minerals that have been altered by water. Although a few shield volcanoes and volcanic domes have been found inside the near-side maria, most of the lava that formed these plains flowed from certain of those ubiquitous lunar features—craters. We will say more about craters in general shortly. The oldest lava flows have been dated at 4.2 billion years, the youngest at 1.2 billion. Most, however, erupted onto the Moon's surface between 3.0 and 3.5 billion years ago.

Contrasting with the maria are the lighter colored *terrae* or *highland* regions. These have been radiometrically dated as being about 4.4 billion years old (so they are significantly older than the maria) and are thought to have formed as a sort of rocky "foam" that built up from the early ocean of magma (technically, these types of igneous rock are known as *cumulates*; in this instance, *plagioclase cumulates*). Although they are often referred to (not altogether inaccurately) as "mountains", it seems that none resulted from the sort of tectonic processes responsible for the mountain ranges of Earth. As for the lunar crust itself, this appears to be about 30 miles (50 km) thick and seems to by mostly composed of anorthosite. Blanketing this is a layer of highly fractured rock many kilometers thick known as the megaregolith and on top of this again, a thin layer of true regolith formed over vast ages as the constant rain of small meteorites and cosmic dust weathers away the surface rock into ever finer particles. The finest particles of silicon dioxide glass have a texture similar to snow and, according to astronauts, smell like spent gunpowder. This is sometimes called "lunar soil" but this is actually a misnomer. Strictly speaking, the term "soil" implies something biologically active. Unlike true terrestrial soil "lunar soil" is (not surprisingly) completely sterile.

Craters of the Moon

Anyone who has looked at the Moon through even the smallest of telescopes cannot help but be impressed by the abundance of one geological feature; craters! Prior to space probes visiting Mercury and Mars, the pockmarked surface of our nearest neighbor presented the only heavily cratered landscape of which we were aware and it was this contrast with the thinly-cratered Earth that made the Moon appear as such an exotic world to we Earthlings. The origin of these features inspired several different theories over the years—some of them plausible and some which were, well, strange! One suggestion saw the craters as analogues of Earth's coral atolls. Another even more way-out idea put forward by Spanish engineer Sixto Ocampo proposed that they were formed by atomic explosions following a fierce nuclear war between two races of (now extinct!) Moon men. During the course of this war, the early lunar seas were blasted into space, some of the water landing on Earth and causing the Noachian flood!

More likely was the hypothesis put forward in the nineteenth century by Franz von Gruithuisen and several other astronomers, namely, that the craters were caused by the impacts of large meteorites. This one eventually won the day but not before being challenged by the equally plausible plutonic or volcanic theory. This model, whose supporters included such noteworthy lunar observers as Gilbert Fielding and Patrick Moore, essentially understood the craters to be the mouths of ancient volcanoes. Moore argued that this better explained the apparently non-random patterns seen amongst the distribution of craters as well as the high percentage of craters whose central peaks are also capped by small craterlets. In addition, he noted that in situations where one crater overlaps another and breaks through its periphery, the one most affected is nearly always the larger of the two. This, he argued, follows naturally if volcanic activity on the Moon declined over a period of time prior to its eventual extinction. Early craters would then tend to be large, but as activity declined, later ones breaking through the same "hot spots" would decrease in size and should also appear fresher and better preserved than the older ones. These latter would tend to appear disrupted by the later volcanic activity—more "broken down" as, indeed, we find them to be in the vast majority of instances. Moore also drew attention to the numbers of large craters which appear to be filled by material suggestive of an upwelling of molten magma that subsequently cooled and solidified. (By the way, he was also favorably disposed toward a volcanic explanation for the craters on Mars' moon Phobos; a far less convincing position to maintain.)

Accumulating knowledge since the days of Apollo has added support to the meteorite theory. Moreover, the lion's share of these impacts have been dated back to the period between 3.8 and 4.1 billion years ago, indicative of a great and sustained influx of objects into the inner Solar System at that time. Just such an influx would be expected according to the Nice Model of the Solar System's genesis, as we saw in Chap. 2.

Nevertheless, the volcanic model was not without some validity. As already mentioned, most of the lava flows that created the maria issued from craters. These are not volcanic vents per se, but the scars of impacts so violent as to puncture the solid crust and allow the escape of molten magma trapped beneath. The few examples of shield volcanoes show that not all volcanic activity resulted from impacts although non-impact volcanism remained only a minor factor in shaping the lunar surface.

Moore's objections to the impact theory can be answered. Apart from the general tendency to see patterns in any set of random points, the impact scenario leaves the door open for genuine patterns to exist in the crater distribution. We now know that it is not uncommon for asteroids to have moons and it is possible that some of the double craters were indeed formed by double asteroids. Even crater chains can be accounted for. Fragile asteroids passing close to Earth could be broken into chains of smaller bodies as Comet Shoemaker-Levy 9 was broken up into a "string of beads" before impacting Jupiter. Moreover, despite the saying that "lightning never strikes twice in the same place" we all know that it does—and meteorites can strike twice in the same place as well. It is thought that larger potential impactors passing through the inner Solar System were more common around four billion years ago and most of the larger craters were probably formed back then. Subsequent impacts would have been smaller, for the most part, accounting for the overlaying of larger craters by smaller ones. Moreover, if a larger object did happen to impact the same site as an earlier smaller one, chances are that the smaller crater would be completely obliterated. For that reason alone, we might expect that, where overlapping craters are found, the smaller one will be the fresher and the larger one older and more seriously degraded. Other objections such as filled-in craters and the existence of craterlets on a fairly high percentage of central peaks (as found by Moore and other observers) can be explained in terms of impact-induced volcanism not unlike that responsible for the lava flows of the maria. Concerning the central-peak craterlets, data from India's *Chandrayaan-1* mission and NASA's *Lunar Reconnaissance Orbiter* published in early 2012 reveal evidence of volcanic vents, lava ponds showing cooling cracks, lava channels and what appear to be volcanic boulders on the central peak of the relatively young crater *Tycho*. This data supports Moore's opinion concerning the volcanic features of central crater peaks, though not the wider view of lunar volcanism championed in his earlier books.

Ice and Atmosphere

The Moon has traditionally been presented as airless and waterless, a complete contrast to its primary, Earth. For all practical purposes this is true but recently some interesting facts started emerging from the continuing acquisition of lunar data.

In 2008, *Chandrayaan-1* found spectroscopic evidence of hydroxyl absorption in sunlight reflected from the Moon's surface. This is a disassociation product of water and its presence in the lunar spectrum betrays the existence of a significant quantity of water ice on the Moon's surface. The following year, *LCROSS* sent an impactor into a permanently shadowed polar crater and detected water in the plume of material ejected by the impact. Then in May of 2011, Erik Hauri and colleagues discovered water in melt inclusions within the sample of volcanic "orange glass soil" brought back by astronauts of the *Apollo 17* mission in 1972. The melt inclusions within this sample formed deep within the Moon some 3.7 billion years ago and the concentration of water within them is comparable with that of the magma within Earth's upper mantle. Although the water in this sample appears to be primordial, that found in shadowed craters and closer to the surface in general was probably deposited there by impacting comets. Its presence may well prove valuable if permanent manned outposts are ever constructed on the Moon.

Regarding atmosphere, outgassing and bombardment of surface material by the solar wind has given our natural satellite a *very* tenuous atmosphere of sodium, potassium, argon-40, radon-222, polonium-210, helium and water vapor. For some as-yet-unexplained reason, oxygen, nitrogen, carbon, hydrogen and magnesium—all of which are present in the regolith and "should" have been released into the atmosphere—seem to be absent there.

Transient Mysteries and Mystery Transients!

With an atmosphere so thin as to be a considered a hard vacuum on Earth and volcanic activity that ended over a billion years ago (save for the occasional localized bout of impact-driven volcanism) the Moon should be a place where very little happens and any changes are few, slow and minor. But there are many lunar observers who assure us that this simply isn't so! Mysteriously, the Moon is not the unchanging place that we might imagine.

For example, way back in 1787, on the night of April 19, an observer as astute as William Herschel spotted three glowing red spots beyond the terminator on the dark part of the Moon. Herschel took them to be volcanoes in eruption, a reasonable enough conclusion at the time. However, we now realize that lunar volcanism has long ceased except for any that might temporarily be induced by a major crater-forming impact. Needless to say, had such an impact occurred in Herschel's day, astronomers of the time would have seen more than three reddish lights! Our planet would have been pelted with lunar debris in the father and mother of all meteor storms, accompanied no doubt by some pieces large enough to strike the ground as meteorites. And, of course, no large crater of extreme youth is located near the region of Herschel's luminous spots.

So what did Herschel see? Who knows? But it may be worthwhile noting that on the same night as his observation, the aurora borealis or northern lights turned on a display that was seen as far south as Italy. This may be pure co-incidence, but it may perhaps hint that the same solar activity triggering the auroral display on Earth also triggered something on the surface of the Moon. On the other hand, there have been even more spectacular aurora without anyone simultaneously finding odd lights on the Moon, although it would be interesting to keep a close watch on the dark region of our satellite whenever a spectacular aurora occurs and the Moon is in a suitable phase.

Herschel was far from alone in his reporting of transient lunar events. Over the years other respected observers spied a wide spectrum of phenomena on the Moon, including spots of color, patches and spots that moved across the lunar surface and occasional obscurations of familiar surface features suggestive of some type of lunar cloud or fog, even though the extremely tenuous atmosphere should be capable of supporting neither. Certain fixed locations on the surface are also the sites of difficult-to-explain changes. The crater *Grimaldi*, for example, has several dark patches on its floor which show variations in hue not obviously consistent with the effects of light and shadow. Noted astronomer W. H. Pickering, who made extensive observations of the Moon between the years 1919 and 1924, noted several dark patches that even altered shape over periods of time, as well as others that moved! Not being a person who feared expressing controversial hypotheses, Pickering opined that these counted as evidence of lunar life; the stationary patches that changed shape and hue suggested fixed organisms analogous to plants while the moving ones were likely swarms of some more mobile type of organism, more or less analogous to Earth's animals or insects. Pickering was quick to point out that the sort of life we know on Earth could not survive for one instant on the Moon but he was nevertheless of the opinion that something which could be called "life" in the broadest sense of that word seemed to be the most likely explanation for what he had seen on our satellite's surface. Few astronomers agreed with him, either then or now. It is difficult to see how any form of life could exist on the Moon and, not surprisingly, there were no hints of any form of biological activity in the samples brought back by the *Apollo* astronauts. Still, Pickering was an experienced observer and he was not alone in seeing moving and changing patches on the Moon. So what is the explanation? Once again, all we can say is "Who knows?"

On the night of November 2, 1958, Russian astronomer Nikolai Kozyrev saw what looked like an eruption on the central peak of the crater *Alphons*. Could some of these central peaks have remained volcanically active millions of years after their formation? This seemed doubtful to say the least, but Kozyrev's observation was to add something extra to the saga of odd lunar events. The telescope he was using at the time was equipped with a spectrometer and he was able to put this to use in obtaining some interesting and controversial data. His first spectrogram appeared to show emission lines of C2 and C3 molecules—not really your typical volcanic effluvia and not something expected on the Moon! While obtaining a second spectrogram, Kozyrev noted that the central region of the crater appeared to brighten and assume an unusual white coloration before fading back to its normal appearance. A second spectrogram taken at that time was quite normal and revealed no indication of gaseous emission. Although spectrographic evidence might seem like proof positive that something was happening here, the first spectrum is apparently not as clear as we might wish and doubts have been raised as to the reality of the alleged emission lines. Be that as it may, Kozyrev noted a second episode at the *Alphons* crater on October 23 the following year. Interestingly, Patrick Moore was observing the same feature on that night and saw nothing unusual, although we do not know if he was looking at *precisely* the same spot at *exactly* the time that Kozyrev noted the brief event taking place. Incidentally, although central peak volcanism is unlikely to have been the cause of the *Alphons* events, it is at least worth mentioning that an earlier event—this time involving the crater *Tycho* (recall that this is the one were evidence of past volcanic activity of the central peak was found in 2012!)—was reported by Walter Haas in the late 1940s. Haas described this phenomenon as "a milky luminosity" but it appears to have been confined to the crater's walls rather than to the central peak.

The reports by Kozyrev plus a multi-colored luminous spot observed at Lowell Observatory by two Aeronautical Chart and Information Center cartographers, James Greenacre and Edward Barr, on the night of October 29, 1963, finally brought transient events on the Moon into the realm of scientific respectability. Nevertheless, scientific respectability does not automatically imply that all scientists are convinced of the reality of these events. Even though a substantial catalogue of *Transient Lunar Phenomena (TLPs)*, as they are now officially known, has been compiled, there are still skeptics who wonder if the whole thing is not simply a case of observational error. Those who support the objective existence of the phenomenon cannot counter skepticism by presenting a watertight account of the mechanism driving these events; at least, not the luminous ones. Residual volcanism of some sort, despite its initial attractiveness as an explanation, is most unlikely when all is taken into consideration. The release of pockets of gas from beneath the lunar surface sounds good initially. Released gas could be induced to glow by solar radiation and the gaseous outrush itself might elevate a quantity of dust adding further to the appearance of a glowing cloud. The biggest problem with this explanation is accounting for the existence of these gas pockets in the first place. If the *Alphons* events of 1958 and 1959 were driven by gas escaping from underground pockets, their gas supply would have needed replacement in just under 1 year. This is not easily explained on current knowledge of the Moon.

On the other hand, the occasional report of obscuring "clouds" or "mist" may have a ready explanation, albeit one that came to light only after yet another lunar mystery was reported. Astronauts on board the command modules of *Apollos* 8, 10, 12 and 17 all noted that lunar sunrise and sunset was accompanied by twilight effects not unlike the crepuscular rays so familiar on Earth. The problem is, effects such as this simply should not be present on a world with so little atmosphere. On Earth, our relatively thick air is easily enough to support particles of dust and droplets of water, as well as refracting the light of the setting Sun and giving rise to what are often spectacular twilight displays. But nothing like this should occur on the Moon. So what did the astronauts see?

The answer was supplied by one of the instruments left behind on the Moon's surface by the *Apollo 17* astronauts. The *Lunar Ejecta and Meteorites (LEM)* instrument was basically designed to monitor dust particles kicked up by meteorites striking the lunar surface, however it soon became apparent that not all the dust being registered came from this source. On the contrary, a significant quantity of dust was found to move across the terminator at slower velocities than that expected for impact ejecta.

Indeed, the terminator turns out to be encased in a diffuse moving dust cloud and it is this "twilight dust storm" that gives rise to the weird effects observed by the astronauts in lunar orbit.

According to Timothy Stubbs of NASA's Goddard Space Flight Center, this dust cloud is generated electrically. The day side of the Moon has been given a positive charge by the effects of solar radiation while the night side becomes negatively charged. At the day/night interface (the lunar terminator) horizontal electric fields are sufficiently strong to both elevate electrostatically charged dust and push it sideways across the terminator. On occasions, we may suggest, portions of this perpetual dust cloud may break off and spread over parts of the daytime side of the Moon, thereby accounting for the occasional report of mist or obscuring cloud. It may also be suggested that differences in polarity similar to that between the day and night side may remain where there are large shadows of ridges or crater walls and that smaller scale dust clouds may therefore be generated quite far from the terminator. Is it possible that sunlight scattered by these transient clouds of dust might even account for the glowing TLPs? One way of testing this would be to check whether reported luminous TLPs occur close to shadows. That would at least alleviate the need to postulate pockets of sub-surface gas periodically blowing out and rapidly being replaced by some unknown mechanism.

In a way, it is good to remember that the Moon—the nearest and most studied astronomical object and the only body other than Earth that emissaries from mankind have visited—still has its mysteries. That alone should keep us from being too presumptuous as to how much we might know about those even more mysterious worlds farther afield.

But one last word about our Moon, before leaving for those worlds "farther afield". Prior to the real Moon landings, fictional accounts of what one may experience on the surface of the Moon sometimes mentioned that shadows there would be completely black; so black indeed that if one were to step into a shadow, one would immediately become invisible to a close companion. Consequently, questions were raised in some places when telecasts of the real Moon landings showed shadows that were not totally dark. This apparent discrepancy has been put forward by conspiracy theorists as one more piece of "evidence" that the *Apollo* program was all enacted in some clandestine film studio (probably in Sector 51 or whatever it is allegedly called—in the room next to the one where the frozen aliens are kept!). It is useless to argue with a conspiracy theorist. Just by raising a counter viewpoint, you become part of the conspiracy and therefore cannot be trusted. But if anyone who has not become committed to the conspiracy cause raises the "problem" of lunar shadows, just remind them that, although there is insufficient lunar atmosphere to *refract* light into shaded areas, there is ample light being *reflected* from the surface to provide some level of illumination within lunar shadows. Shadows on the Moon should not be totally dark after all. There is no need to invent a conspiracy, just to recognize the presence of reflection!

"The Queer Little Moons of Mars"

Apologies to the memory of V. A. Firsoff for this quote, but it was the title of a chapter on these objects in his book about the planet Mars. The primary dictionary definition of "queer" is "odd or eccentric" and no better term could characterize these strange little objects.

As if to set the scene of strangeness, a pair of small Martian moons was twice proposed in fictional literature before they were discovered in fact; In Jonathan Swift's *Gulliver's Travels* (1726) and again in Voltaire's short story *Micromegas* (1750). Discovery of the genuine articles did not, however, take place until 1877 when Asaph Hall of the US Naval Observatory in Washington DC found the fainter and outermost moon on August 12 and the brighter, but innermost, on August 18 during the course of a deliberate search for possible satellites of the Red Planet. Subsequently, the moons were named after the mythical sons of Ares (Mars), the god of war, whom they accompanied into battle. The outermost moon was named Deimos (terror or dread) and the innermost Phobos (panic or fear); suitable accompaniments of war. The rather late discovery of these moons was due, not so much to the faintness of the objects themselves as to the overpowering brightness of their primary. The smaller and fainter Deimos has an average magnitude of about 12.7 and Phobos 11.6. Alone in a dark sky they should both be within range of a 10- in. telescope but, needless to say, Mars makes sure that they are *not* alone in a dark sky! Patrick Moore managed to find both while using a 15-in. reflector and on at least one occasion caught sight of Phobos in a 12.5-in. telescope. He opines that very keen-eyed observers might be able to track the pair down with a 12-in. if conditions—and telescope—are first class.

> **Project 1: Can You See the Moons of Mars?**
>
> Neither moon of Mars would require an especially large telescope if located in a dark part of the sky. A 10-in. (25.4-cm) reflector should be enough to see both of them under these conditions; Phobos quite readily and Deimos as a somewhat more difficult object. However, neither *is* visible in a dark sky. That's the problem with moons of other planets—by their very nature there is this bright object called the primary planet forever spoiling the view!
>
> Nevertheless, each moon has been observed in relatively small telescopes where conditions and optics—as well as the eyesight of the observer—are all first class. As we saw, a very experienced observer like Patrick Moore managed to see both in a 15-in. and Phobos in a 12.5-in. The reader might like to try his/her hand at tracking down these little moons. You will need a good set of positions for the night of your attempt and a good steady sky. An eyepiece with an occulting bar to block out Mars will be a real advantage. If you own, or have access to, a telescope of 12 in. (30 cm) or larger and are successful in locating one or both moons, you may try reducing the aperture by placing opaque cardboard sheets, with holes of different diameters cut in them, over the aperture of the telescope in order to find the smallest effective diameter that allows the moons to remain visible. Is it possible to see them using smaller apertures than 12 in.?

Compared with Earth's giant Moon, the satellites of Mars are very small fry indeed. The larger of the pair—Phobos—has an average diameter of 13.9 miles (22.2 km) and Deimos a mere 7.9 Miles or 12.6 km. Like most small bodies, both moons are not spherical but are rather more elongated in shape. Phobos, the innermost moon, orbits just 5,861 miles (9,377 km) from the center of Mars

190 Weird Worlds

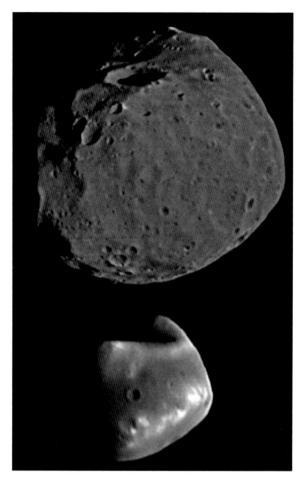

FIG. 4.2 The Martian moons Phobos (*top*) and Deimos (Credit: NASA)

(or, put another way, a mere 3,800 miles, in rounded figures, from the Martian surface!) and Deimos a somewhat more respectable 14,663 miles (23,460 km) from the center of the planet (Fig. 4.2).

The extreme proximity of Phobos to the surface of Mars would have some odd consequences for a Martian observer. Because Phobos (and Deimos too for that matter) orbits the planet pretty much in the equatorial plane, the best location for Phobos watching is at the equator. But because the moon whips around Mars in just 7.66 h—faster than the planet itself rotates—it rises in the west and speeds across the sky to set in the east just four and a quarter hours later. During that time period, it progresses through more than half its cycle of phases from "new" to "full". Following this dash across the Martian sky the moon remains out of sight for six and three quarter hours before again appearing in the west 11.1 h after its previous time of rising. Except for brief periods near both mid summer and mid winter, Phobos does not complete one circuit of the sky without passing through Mars' shadow at some point. It has been calculated that Phobos is eclipsed some 1,330 times during each Martian year!

The moon is closest to the hypothetical equatorial observer when directly overhead and at that time appears about 12 minutes of arc in diameter or somewhat less than a third the diameter of our own Moon. However, at moonrise and moonset the distance of Phobos from the observer is so much greater that it subtends a diameter of only 8 minutes of arc! For observers on Earth, it is well known that the rising and setting Moon *appears* larger than when seen high in the sky; a phenomenon known as the Moon illusion and one known to be a perceptual effect only, not involving any true change in the measured diameter of the Moon's image. On Mars, something like an inverse Moon illusion occurs—except that in this instance it is *not* an illusion. The measured diameter of Phobos really *is* less near moonrise. (The present writer hopes that when humans eventually land on Mars, someone takes the time to observe both the actual and perceived size of Phobos when rising and when high in the Martian sky to see how—or *if*—the Moon illusion works with something that really *does* change in apparent size, albeit in a direction opposite to that of the illusion. Will the Moon illusion cancel out the perception of the real effect, such that Phobos will *appear* to remain about the same size? If any reader of this book ever winds up on Mars, please consider this as a further Project and carefully note the results!). A similar effect would be noted as one draws away from the equator of Mars. The angular diameter of Phobos gets smaller the further that one ventures from the equator. The moon also appears lower and lower in the sky. Above 69° latitude, north or south, the moon remains permanently below the horizon and forever invisible.

At its brightest, as seen from the surface of Mars, Phobos appears about equal in its total light to Venus as seen from Earth. Nevertheless, because that light emanates from a larger angular area, its intensity is not as great as that of Venus herself.

Because Phobos orbits Mars faster than the planet rotates, its orbit is slowly decreasing and the moon edging ever nearer to its primary. About 40 million years from now, it approaches the planet so closely that tidal forces will literally tear it to shreds; the debris probably forming a ring around Mars, at least in the short term. Eventually this too decays and the surface of Mars will be struck by pieces of its former moon. It is thought by some planetary scientists that similar events have happened in the past and left evidence in the chains of impact craters that have been noted on the planet. Maybe Phobos is simply the last of a population of small moons that have spiraled into the planet over the ages.

Recent close-up measurements of Phobos have found that it has a surprisingly low density; just 1.88 times that of water! This is lower than the least dense meteorites recovered on Earth and hints at the existence of voids within the body of the moon. It is possible that these voids and pores were once filled, at least partially, with ice. Although the presence of impact craters has been revealed by space probes, certain pit-like features on the moon's surface appear to have a different origin and may be more in the nature of vents where water vapor has jetted forth at times in the past. Some ice may exist there even today and Russian spacecraft have found evidence of a low level of gaseous emission, possibly indicating the contemporary presence of sublimating volatiles within the body of Phobos.

Located further from the planet, Deimos takes 30 h to complete an orbit of Mars, a little longer than the time it takes for Mars to make one full turn about its axis. This means that, unlike Phobos, Deimos rises in the east in the conventional manner, but because the rotation of the planet almost keeps pace with its drift across the sky, it takes an inordinately long time to make it across to the western horizon. Indeed, although it orbits the planet once in 30 h, it takes *60 h* to cross the Martian sky from its rising in the east to its setting in the west, followed by another 60 h gap until its next rising. As it crosses the sky, it goes through its entire cycle of phases *twice*, although its changes in apparent shape would not be conspicuous without optical aid. Being so small and relatively distant, the angular diameter of this moon is only about 2 minutes of arc as seen from the surface of Mars. This is not much larger than that of Venus as observed from Earth. From Mars, Deimos appears little different to any star or planet except for its slow movement against the background sky and its variations in brightness as it goes through the cycles of phases. Like its fellow satellite, Deimos is more favorably placed from the Martian equator and, again like Phobos, never rises above the horizon at polar latitudes. Because of its greater distance however, a larger swathe of higher latitudes have the chance of seeing it and it does not remain permanently below the horizon until latitudes 82°, north or south, are reached. It is also subject to eclipse in Mars' shadow, with these events occurring some 130 times each year. Both Phobos and Deimos share with our own Moon a tidal locking which means that each has only one of its faces perpetually turned toward its primary.

From Mars, the Sun subtends an angle of 21 minutes of arc. As this is considerably larger than the maximum of 12 min of Phobos, not to mention the 2 min of Deimos, it is readily appreciated that total solar eclipses never occur on the surface of the planet. Nevertheless, non-total solar eclipses do occur quite frequently. Phobos is responsible for 1,300 eclipses every year, but they are extremely brief as the moon crosses the solar disk in just 19 s. Deimos crosses the Sun's disk 120 times per year, taking 2 min to pass right across. It is debatable whether the Deimos events should really be called eclipses or transits, as the moon only covers one ninth of the solar disk. This raises the question of where an eclipse ends and a transit begins, but that is one that this writer is very happy to leave to somebody else!

It is interesting to note that both moons can cross the solar disk together, and it seems that these double eclipses occur quite often. Future astronauts will no doubt be awed by this literally un-Earthly spectacle. Not as impressive although still interesting, as well as being unlike anything seen on Earth, mutual eclipses can also take place as Phobos eclipses Deimos. Once again, definitions might raise their ugly heads here as the angular size of Deimos is almost small enough to allow such events to be described as occultations but, as above, sorting out this terminology issue can be left to others.

Whence Came the Martian Moons?

Until recently, it was almost universally accepted that both Martian moons were originally asteroids captured by the planet. This sounds fine when expressed in broad terms, but getting down to the details of how this could have happened has not been at all straightforward. As a small planet, Mars is naturally a lot less capable of capturing asteroids than, say, Jupiter where many of the objects now classified as moons are almost certainly captured asteroids. But even the asteroid-moons of Jupiter usually betray themselves by having eccentric orbits. It is not easy to get a captured body into the essentially circular orbits of the moons of Mars. Moreover, both orbit almost precisely in the planet's equatorial plane, a display of regularity not altogether expected from captured objects. Attempts to get around this by invoking the drag of an extended and far thicker atmosphere than that possessed by the Mars of today lack any real supporting evidence. The capture of a binary asteroid has also been proposed; the two objects separating to become Martian moons following initial capture. The twin problems of the eccentricity and low inclination of the moons' orbits, however, remain.

Although broadly classifiable with C-Type or carbonaceous asteroids, the moons do not fit easily into any strict asteroid type, nor are they close analogues of any known meteorite. The low density and high porosity of Phobos also casts doubt on an asteroidal origin. Moreover, Phobos seems to be largely composed of phyllosilicates, which are also common on the surface of Mars itself. This suggests that the Martian moons may have formed in a way not unlike that proposed for Earth's Moon, i.e. through an early collision between Mars and a large body. Material ejected from Mars following a major impact may have coalesced to form the present moons, and possibly a population of others that have since spiraled back onto the planet itself. That would account for the apparent similarity between the composition of Phobos and the surface of Mars. On the other hand, if it is true that Phobos does—or one did—contain a significant quantity of ice, an origin depending upon accretion of debris generated by a large impact seems less likely. Maybe the Martian moons formed in situ together with their primary, maybe from a ring of material surrounding the infant Mars, although it must be said that this has never been a popular hypothesis. Future research will undoubtedly cast more light upon this topic and hopefully lead to a true account of the genesis of these two remarkable objects.

Interestingly, the low density of Phobos and the seeming lack of similarity between this moon and any asteroid or meteorite type casts doubt on a suggestion made several years ago that the Kaidan meteorite—a CR2 carbonaceous chondrite that fell in Yemen on December 3, 1980—might have originated as a fragment of Phobos. The present writer has suggested that this meteorite may in fact be a member of the Geminid meteor stream and therefore a fragment, not of Phobos, but of the asteroid 3200 Phaethon. Further discussion of this will not be pursued here however as the topic has already been covered in my former book *Weird Weather: Tales of Astronomical and Atmospheric Anomalies* and can be followed there by interested parties.

The Moons of Jupiter

Jupiter, the king of the Sun's planetary companions, is suitably bejeweled with a large array of moons. Nobody knows what the final count might be as new ones continue to be discovered, but as at mid 2012, the count stood at 67. That will almost certainly go higher and probably will *be* higher by the time you read these words. Most of these moons are, however, *irregular satellites*; in other words, objects that did not form in their present locations as moons orbiting the planet. It is thought that they were originally asteroids captured by the giant planet in its youth. These irregular moons are all small and orbit their host world at large distances in an array of orbits having a range of eccentricities and inclinations to the Jovian equator. The irregular satellite system is therefore rather messy, with some of the satellites even moving in retrograde orbits—"going backwards" from the point of view of their better behaved siblings. The outermost moon discovered to date (thus far, simply known as S/2003 J2) is almost as far from Jupiter as Venus, at its closest approach, is from Earth and take almost three terrestrial years to make just one orbit of the planet!

What's in a Name?

Little more will be said about these small satellites and, indeed, few individual details are known about them. Most likely, their composition is similar to that of the outermost asteroids or cometary nuclei; mixtures of ice and rock with a fair helping of reddish-dark organic material.

The most recent discoveries are still known by number designation rather than by name. Indeed, it took an unusually long time for the brighter members of this class (those known prior to 1960) to receive proper names, to the frustration of astronomers such as Patrick Moore. In his writings in the 1960s, Moore adopted several unofficial names for the fainter moons of Jupiter; names which had earlier been suggested by B. G. Marsden but for some reason never officially adopted by the astronomical community. These have now been replaced, but in case one comes across the earlier monikers, the writer takes the liberty of listing the old and new "official" names, together with the official numbers, of these satellites. The numbers and official names are on the left with Marsden's suggested (but now discarded) name following in brackets. They are;

VI Himalia (Hestia)
VII Elara (Hera)
VIII Pasiphae (Poseidon)
IX Sinope (Hades)
X Lysithea (Demeter)
XI Carme (Pan)
XII Ananke (Adrastea)

Regular Moons

By contrast with the above, the regular satellites of Jupiter all orbit very sedately in almost circular orbits having only small inclination to the equatorial plane of the planet. Contrasting again with the large number of irregular moons, the regular ones are relatively few in number—just eight in total; the four Galilean moons, plus Amalthea, Metis, Adrastea and Thebe. Of these, the famous four "Galilean moons" are huge. These are Io, Europa, Ganymede and Callisto ranging in diameter from 1,961 miles (3,138 km) for

Europa to 3,289 miles (5,262 km) for giant Ganymede. This quartet is outstanding; the rest of Jupiter's moons (be they regular or irregular) are all less than 125 miles (200 km) in diameter (although the elongated Amalthea has a long axis of nearly 169 miles or 270 km). Indeed the majority have diameters of only about 3 miles or 5 km.

The four large and four small regular moons orbit relatively close to the primary with orbital radii ranging from just under 80,000 miles (128,000 km) for Metis to 1,175,000 miles (1,880,000 km) for Callisto. Metis and its neighboring moon Adrastea are so close to Jupiter that they whip around the giant planet in just over 7 h, a shorter time than Jupiter itself requires to make one full rotation about its axis (contrast this with the period of nearly 3 years for the outermost of the irregular moons). By the way, this Adrastea is not to be confused with the Marsden/Moore Adrastea; the irregular moon now known as Ananke.

Unlike the captured irregular moons, these eight satellites formed together with Jupiter itself in what may be regarded as a miniature version of the pre-solar nebula. It is thought that many more moons similar to the "big four" Galilean satellites accreted from this nebula, but like planets forming around stars, most of these migrated inward and eventually plunged into the planet itself. This process of formation, migration and destruction continued until the matter in the nebulous disk became too thin to seriously perturb the motion of the satellites and the process halted. On this model, the remaining regular moons are understood as being the final ones to form; the ones fortunate enough to have exhausted most of the remaining disk material in their accretion.

The Moons of Galileo

The so-called "big four" or Galilean moons are not just the largest but also the most interesting of Jupiter's retinue. Historically, they are interesting because Galileo's discovery that they orbited Jupiter effectively disproved the prevailing cosmological model of a geocentric Universe. They are also among the first things tracked down by a novice astronomer and, together with craters on the Moon and the rings of Saturn, inevitably prove a hit when the neighbors come to have a look through your telescope. They are also quite bright and easy to see in the smallest of telescopes. Ganymede would be a relatively easy naked-eye object from dark sky locations were it not for the presence of brilliant Jupiter. Even so, keen-sighted folk have indeed managed to see it sans optical aid. Indeed, a few people of exceptionally sensitive eyesight have claimed detection of *all four* Galilean moons with their naked eyes.

> **Project 2: Seeking Ganymede with the Naked Eye**
>
> The magnitude of Ganymede has been estimated as bright as 4.6, which would normally make it a relatively easy naked-eye object even in less than perfect skies. However, the very nearby Jupiter shines some 620 times brighter than its largest moon, rendering the fainter object a lot more difficult than it otherwise would be. Naked-eye sightings have nevertheless been reported and if you are blessed with very good eyesight, just such an observation may be possible for you. The best way is to locate the moon first through binoculars and then try for it without optical aid. Eyesight free from astigmatism is probably necessary for success as astigmatic vision causes the image of Jupiter to flare, making it very difficult to locate a fainter object nearby.
>
> If you manage to find Ganymede quite readily without optical aid, you might like to try the harder task of finding other Galilean moons with the naked eye. There have been, as mentioned, credible accounts of all four having been spied in this way, though this is a feat not often accomplished and surely requires exceptional vision together with perfect sky conditions.

Yet, for all that, little was known about the moons as individual worlds. They may look like bright stars in small telescopes, but large instruments do not give a greatly clearer view. Their disks remain tiny and although some very experienced visual astronomers observing from locations such as the Pic du Midi Observatory in the French Pyrenees recorded light and dark markings on these diminutive disks, their significance was not easy to determine. One of the few things that could be confirmed was that, unlike

Fig. 4.3 Comparative sizes of the Galilean moons (Credit: NASA)

Saturn's giant moon Titan, these Jovian satellites did not possess any substantial atmosphere. In the main, they tended to be dismissed as large, but not especially interesting, balls of airless rock. How wrong we were! The first space probes were to prove that these relatively minor bodies are amongst the most interesting—and in many ways the weirdest—objects in the entire Solar System! (Fig. 4.3)

Before moving on to consider each of the Galilean moons in turn, it is interesting to note that the first three, Io, Europa and Ganymede, appear locked in a peculiar dance known as a Laplace resonance. For every orbit of Ganymede, Europa goes twice around Jupiter and Io four times, but in such a way that when Europa and Io are on opposite sides of Jupiter, Io is at the point of its orbit that lies closest and Europa at the point of its orbit lying furthest from the planet. However, when Europa and Ganymede lie on opposite sides of Jupiter, the former is at its closest to the planet. Because the longitudes of Io—Europa conjunctions and Europa—Ganymede conjunctions change at the same rate, a conjunction of all three moons is not possible; there cannot be a time when one of the moons is directly on the other side of Jupiter from the remaining two. This complex resonance does not result in serious pumping up of the moon's orbits to high eccentricity, unlike the resonances that we noted in the asteroid belt.

Io

Travelling outward from Jupiter, the first of the Galilean moons one would arrive at is also the weirdest; Io. Orbiting its massive primary at a distance of 262,000 miles (421,700 km) from the planet's center (or just 217,000 miles or 350,000 km from the Jovian cloud tops), Io takes a mere 42.5 h to whip around the giant world in an orbit which—though hardly very eccentric by the standards of many bodies in our planetary system—departs from circularity further than we should normally expect for a body so close to a mass as large as Jupiter. Its orbital eccentricity is 0.0041, not a very great departure from circularity we might think, yet one that requires explanation and which also has some fascinating consequences. We will look at both in a little while.

Because of its proximity to Jupiter, Io (and, indeed, all of the Galilean moons) have their orbits locked in to Jupiter's rotation and in common with Earth's Moon, have one face perpetually turned toward the primary. With respect to its size, Io is the second smallest of the Galilean moons, at 2,282 miles (3,652 km) diameter, being just a little larger than Earth's satellite. Having a density of 3.5275 times that of water, it is also a little denser than our Moon and actually holds the prize for being the densest moon in the Solar System, making it a real oddity amongst the principally icy major satellites of the outer planets. Contrasting with these objects, Io is thought to be composed principally of silicate rock and iron and to possess a structure more like that of a terrestrial planet than an outer Solar System moon. That is to say, it probably has a core made of either iron or iron sulfide accounting for around 20 % of its entire mass. If this is essentially iron, it is estimated to be between 440 and 800 miles (700–1,300 km) in diameter. If it is a sulfur/iron mix, it would be somewhat larger; between an estimated 680 and 1,120 miles (1,100–1,800 km) across. Either way, the core is thought to be surrounded by a silicate mantle, around 10 % of which consists of an ocean of molten magma some 31 miles (50 km) deep and sporting temperatures about 1,200 °C. This magma "ocean" is trapped like a sandwich layer about the same depth beneath the moon's surface. Above it is a lithosphere of basalt and sulfur between 7 and 25 miles (12–40 km thick). On the surface rise clusters of seriously high mountains. Their average height is around 4 miles or 6 km with some rising to a full 11 miles (17 km) above the mean surface. These are higher than Earth's Mt. Everest, and considering how much smaller Io is than our planet, their comparative height is truly awesome. But neither its towering mountains nor its unusual composition is the real feature that sets Io apart. There is something even more dramatic about this moon!

The first hint of something truly weird about this moon came in 1979 when *Voyager* navigation engineer Linda A. Morabito noticed a mushroom-shaped plume rising over its limb in the *Voyager* images. The general form of this feature bore a remarkable resemblance to speculative paintings of how a major volcanic eruption on our own Moon would look, if such events did truly occur there. Subsequent observation of the Io plumes proved that this is precisely what the plume was. Far from being the relatively uninteresting ball of rock that Io was once supposed to be, the moon turned out to be the Solar System's Yellowstone Park; the most volcanically active body yet discovered!

Images of the surface of Io reveal a multiplicity of volcanic vents and lava flows. Someone described the multi-colored face of the moon as looking like a pizza, and there is truth in that description. Impact craters are conspicuous only by their absence, indicating a rapidly recycling and highly active surface. Sulfur and silicate flows make colorful patterns. Red deposits left by explosive volcanism are thought to be composed of three and four chain molecular sulfur while the more familiar forms of this element leave characteristic yellow deposits. Sulfur dioxide gas venting from the volcanoes freezes on the surface amid flows of searing hot basaltic lava. It was once thought that the lava was mostly sulfur, but more recent temperature measurements show that it is too hot to be composed of this element. Like terrestrial lava, it is basaltic with, it seems, a high magnesium content. The sudden release of sulfur and sulfur dioxide from erupting magma at volcanic vents and lava lakes blasts plumes to heights of over 180 miles (over 300 km) into the sky, the material falling back to the surface over such large distances that rings in excess of 600 miles (1,000 km) in diameter encircle the vents. Smaller plumes—though still erupting to heights of 60 miles (100 km) or thereabouts—occur when flows of molten lava spread across layers of sulfur dioxide frost setting off explosive sublimation that blows the lava sky high. Although not quite as dramatic as the really big ones, these plumes tend to be longer lasting.

The big question is "Why is this small world so excessively volcanic?" We might suppose that the internal heating from the decay of radioactive materials such as drives volcanism on Earth should have long since dissipated. And we would be right! It is not the decay of radioactive elements that drives Io's volcanism. To see what really is at work here, we must recall our earlier remark about the eccentricity of the moon's orbit. But what connection could there possibly be between Io's orbit and volcanism? Let's look at this a little more closely.

Io, as already stated, is the innermost of Jupiter's four large "Galilean" moons. Beyond it lie Europa, Ganymede and Callisto and it so happens that Io orbits in resonance with the former two of these. It goes twice around its orbit as Europa traverses its orbit once and four times for every single orbit or Ganymede. Expressed more formally, it is in 2:1 resonance with Europa and 4:1 with Ganymede. The regular tug of these two moons tries to pull Io outward and make its orbit more eccentric. But at the same time, the constant pull of Jupiter tries to make its orbit more nearly circular. So poor old Io is being pulled in two directions at once by this planetary tug-of-war! Its orbit is forced to be somewhat more eccentric than it would have been had the other moons not been present but, more dramatically, this struggle heats the moon's interior and it is this tidal heating that causes and maintains the high level of volcanic activity.

In the course of our discussion of Jupiter in Chap. 2, we saw how material erupted from Io interacts with the giant planet's magnetic field to produce the most powerful radio beacon in the Solar System. It is estimated that Jupiter's magnetic field sweeps up gas and dust expelled from the moon at the rate of 1 t each second. Most of this material consists of ionized sulfur, oxygen and chlorine, together with atomic sodium and potassium, molecular sulfur dioxide, sulfur and sodium chloride dust. Extending out to around 6,800 miles (nearly 11,000 km) from the moon, is a cloud of neutral sulfur, oxygen, sodium and potassium atoms, some of which end up in orbit around Jupiter itself. Over a period of 20 h, these particles spread out from the moon into a banana-shaped cloud of neutral atoms that reaches over a quarter of a million miles (more than 420,000 km) from Io, both in front of and behind the moon. As we saw in Chap. 2, Io orbits within a doughnut-shaped plasma torus of sulfur, oxygen, sodium and chlorine ions. The immediate source of this is the neutral atomic cloud.

As this impinges upon the Jovian magnetosphere, the neutral atoms of the cloud are ionized and carried along by Jupiter's magnetic field. The ions within the torus rotate together with Jupiter's magnetosphere, being swept along in their orbit of the giant planet at a speed of 46 miles (74 km) per second. Because the Jovian magnetic field is tilted with respect to Jupiter's rotation axis, the torus is also tilted with respect to the planet's equator. As Io orbits within Jupiter's equatorial plane, this discrepancy means that Io is not always within the plasma torus that it created. Sometimes it is above and at other times it is below it.

In 1992, the *Ulysses* spacecraft recorded a curious phenomenon. Particles of sodium chloride—table salt—having dimensions of just 10 μm were detected streaming away from the environment of Jupiter at velocities of several hundred kilometers per second. It is pretty obvious that the cosmic salt shaker is Io, but just how these salt particles are ejected at such high speeds remains a mystery. Are they spat out by volcanic activity or are they blasted off the moon's surface by the intense radiation of particles within Jupiter's magnetosphere, or by some other means? At present, we do not know.

As briefly mentioned in Chap. 2 of this book, Jupiter's magnetic field lines couples Io's very thin atmosphere (more will be said about this later) and the neutral atomic cloud, to the upper atmosphere at the planet's magnetic poles. This generates an electric current as high as two trillion watts known as the *Io flux tube*. It is this that is responsible for auroral glows in both Jupiter's polar regions and even in Io's very tenuous atmosphere. The bright auroral spot produced in this way on Jupiter is known as *Io's footprint* and has a strong influence on the radio emissions from this planet. Jovian radio emissions picked up from Earth are strongest when Io is also visible. Jovian magnetic field lines that go through Io's atmosphere to the moon's surface induce a magnetic field within its interior, probably within the partially molten silicate magma ocean spoken of earlier. This, by the way, is the only magnetic field detected on the moon; no such field is associated with Io's core.

The reader will note that on a couple of occasions during the above paragraphs, mention has been made of Io's atmosphere. Earlier, it was said that the Galilean moons had no substantial atmosphere. There is no contradiction. The atmosphere of Io is hardly rated as "substantial" in the sense that it does not support what we might call "weather" or generate clouds that obscure the moon's surface. The "atmospheric pressure" on Io varies between about 3.3×10^{-10} and 9.9×10^{-14} times that of Earth's sea level. On Earth, it would be classed as a vacuum, but it is still enough to give rise to auroral glows and to interact with the magnetic field of Jupiter. It consists for the most part of sulfur dioxide, with minimal amounts of sulfur monoxide, sodium chloride and some atomic sulfur and oxygen. Thanks to the magnetic field of Jupiter, it is continually being stripped away at the rate of about one ton each second, so it must be continually replenished by the ever-present volcanism, plus some contribution from the evaporation of surface sulfur dioxide frost. When the temperature drops—for instance during the night or when Io passes into the shadow of Jupiter—much of the atmosphere freezes out on the moon's surface. This is less pronounced during eclipses as a layer of sulfur monoxide close to the moon's surface slows the process somewhat during the relatively brief time of an eclipse, but on the nigh side of Io there is enough time for the surface pressure to decrease from 100 to as much as 10,000 times that of noon.

As stated earlier, auroras caused, not by the solar wind as on Earth, but by particles travelling down the lines of Jupiter's magnetic field, occur in the atmosphere of Io. High resolution images from Earth showing the moon passing in eclipse through Jupiter's shadow have detected these and find that they are brightest near the moon's equator. This is, of course, the opposite of terrestrial aurora, but because Io has no magnetic field generated within its core, there is no reason to expect the occurrence of polar aurora there. Close to the equator, the lines of Jupiter's magnetic field are more nearly tangential to the moon's surface and therefore must travel through a greater expanse of atmosphere than they encounter at higher latitudes. Auroras are brighter there simply because the orientation of the field increases the depth of atmosphere through which charged particles swept along by these field lines must pass.

Altogether, Io turns out to be a strange and fascinating little world, albeit one best studied from afar! Somehow, I cannot imagine any future entrepreneurial enterprise building a space-age resort there. With volcanoes shooting torrents of lava 300 miles high, raining down over hundreds of miles of surface, fountains exploding skyward as molten lava floods across freezing sulfur dioxide deposits and the whole place bathed in lethal radiation, Io sounds more like the abode of the damned than a destination for space-age honeymooners!

Europa

Europa is the second Galilean moon, and the sixth of all the Jovian moons, from the planet, orbiting it at a distance of 419,313 miles (670,900 km) with a period of 3.55 days. Having a diameter of 1,961 miles (3,138 km), it is the smallest of the Galilean "Big Four", being just a little smaller than Earth's Moon. Even so, it still rates as the sixth largest in the Solar System.

Like Io, albeit to a far milder degree, Europa is the victim of a tug-of-war between giant Jupiter and the other Galilean moons. Its orbit departs a little from circularity (eccentricity = 0.009) thanks to disturbances by the other Galilean moons; sufficient to cause a tidal flexing as the distance from Jupiter changes slightly during the moon's revolution around the planet. This tidal flexing heats the interior of Europa, as a similar effect heats the interior of Io.

Europa is thought to be made up primarily of silicate rock, not unlike the four innermost of the Sun's planets. There is most likely a metallic iron core, overlain by a silicate mantle which in its turn is surrounded by a layer of water around 62 miles (100 km) deep. The surface of this layer is frozen, and it is this shell or crust of ice that we see as the moon's surface. This icy shell is probably about 6–19 miles (10–30 km) thick and presents a surface which rates as one of the smoothest of any known body in the Solar System. Unlike other planetary bodies, Europa sports very few impact craters but instead is crisscrossed by a patchwork of lines that give it the appearance of (as someone graphically remarked) a "cracked egg". These appear to be lines of coloration rather than significantly elevated features. We will return to these, and other, surface features in a little while.

Although the surface is frozen solid, it is not immediately obvious what lies between this ice shell and the rocky mantle of the moon. As stated earlier, it is water, but is it truly in liquid form or is it better described as "slush" or soft ice?

There are good reasons for believing that this layer of water truly is in a liquid state and may justifiably be described as an "ocean." Tidal flexing probably supplies enough heat to maintain an ocean of liquid water, especially if its salt content is high. The suggestion has also been made that Jupiter might cause large, slow-moving, tidal waves in the moon's ocean capable of dissipating energy into this layer and thereby helping to keep it in a liquid state. Be that as it may, evidence that the ocean is indeed both liquid and salty comes from *Galileo* magnetic field measurements. These suggest that interaction with the magnetic field of Jupiter induces a magnetic field in Europa, the presence of which is best explained by a layer of conductive material beneath the moon's surface. Liquid salt water is a good candidate for such a layer. Additionally, there have been suggestions that the subsurface ocean may be quite acidic and this also would assist its electrical conductivity. There are further indications that the surface has undergone a shift of around 80° over time, implying a certain detachment of the surface from the deeper mantle; something that would be unlikely if the water layer was frozen all the way down to the mantle. Evidence for this shift comes from a closer examination of the moon's surface markings.

Concerning these peculiar markings on the moon's icy surface, the most striking are the dark streaks that give the moon its (badly!) cracked-egg appearance. The larger streaks are over 12 miles (20 km) wide and sometimes have dark and diffuse edges as well as regular patterns of striations and lighter central bands. Their predominant color is reddish. These features are simply known as *lineae* (lines) and close examination of images reveal the edges of the icy crust on each side of them to have moved relative to one another. This suggests that the lineae mark sites of eruptions of warmer ice as the crust fractured and spread to expose the warmer layers under the surface. This fracturing most probably resulted from the tidal stresses caused by Jupiter. Being tidally locked to the planet, Europa always maintains the same approximate orientation to Jupiter and this should result in a very distinctive and predictable pattern of fractures in the ice. The more recent ones do indeed conform, but the older ones do not. Moreover, the older the lineae appear to be, the further they diverge from prediction. This is why scientists think that Europa's surface is shifting. In short, the patterns can be explained if Jupiter's pull on the moon's icy crust causes this to rotate slightly faster than the interior. This, in turn, implies that the surface is decoupled from the moon's rocky interior; in some sense, it "floats". It is this Europan oddity that has convinced most planetary scientists that at least some of the sub-surface water exists in the form of a liquid ocean rather than as a deep layer of slushy ice all the way down to the rocky mantle. By comparing *Voyager* and *Galileo* images, it is estimated that a full turn of the outer shell of ice takes at least 12,000 years to complete.

Another set of features noted on the moon are small circular and elliptical *lenticulae* (freckles). These appear to come in a variety of forms; some seem to be domes, others pits and yet others simply dark spots on the surface ice. One hypothesis suggests that they are caused by upwelling warmer ice pushing through the colder surface crust. This could push up dome-shaped features. The smooth dark spots might form when the warmer ice bursts through the surface and melts.

Alternatively, the lenticulae might simply be small and poorly observed examples of another kind of Europan feature—*chaos terrain*. These "chaos" regions appear to be where warmer ice from below has risen up through the icy crust in a sort of low-temperature analogue of molten magma rising through the solid rocky crust of Earth. Small fragments of the icy crust would be jumbled together, embedded in material that had risen up from beneath. However, some workers in the field interpret the chaos regions as places where the subsurface ocean has melted through the icy crust. This interpretation depends on a controversial minority opinion that the ice crust of Europa is relatively thin, maybe only a few 100 ft in thickness. If this model is correct, it would at least make access to the subsurface ocean easier for future space probes. However it has one serious defect. Although impact craters on Europa are rare, some have been found and the largest examples are surrounded by concentric rings and filled with relatively flat, fresh ice. The colliding body seems to have slammed into thick ice and blasted out the crater rather than penetrating through a thin surface layer and ending up in the ocean beneath. Moreover, calculations of the heat generated by tides predicts a thick ice layer comprising a "cold" and rock-hard outer crust and a layer of "warm" softer ice between this and the liquid ocean. If tidal heating is enough to melt a thin layer, all the water would escape into space and a very different Europa would have greeted the robotic eyes of *Galileo*!

Late in 2011, evidence was presented in the journal *Nature* suggesting that chaos terrain may be related to enclosed "lakes" of water existing rather like bubbles of liquid within the ice shell itself. These lakes, if they exist, would be isolated from the ocean and much closer to the surface. In a sense, this suggested explanation for the chaos terrain proposes a localized thin ice model whilst retaining thick ice for the Europan crust in general.

Differences in color across the surface of Europa apparently result from the differing mineral content of the ice in different localities. Spectroscopic analysis suggests that the reddish material marking the lineae might be rich in magnesium sulfate and other salts deposited as the water erupting from below evaporates in the near vacuum at the moon's surface. Alternately, sulfuric acid hydrate has been proposed. Depending upon which (if either!) interpretation proves correct, the ocean of Europa seems to be either one of brine or of dilute sulfuric acid—or maybe a mixture of both. The sulfuric acid interpretation may have some support by the fact that although both sulfuric acid hydrate and magnesium sulfate are colorless, the reddish appearance of the lineae means that some colorful material must be mixed together with them. Sulfur, similar to the red forms of this element found on Io, is a likely candidate and, if sulfur is present in the subsurface ocean, sulfuric acid may well be there as well.

Being exposed to space, the icy surface of Europa is constantly being bombarded by ultraviolet radiation from the Sun and ions and electrons from the magnetosphere of Jupiter and the combined effect of this is to split molecules of water in the ice into their constituents of hydrogen and oxygen. Hydrogen quickly escapes into surrounding space, where it forms a ring of gas in the vicinity of the moon's orbit, eventually becoming ionized and joining Jupiter's plasma. Some of the oxygen ends up there as well, but some is also retained in a very tenuous atmosphere with a surface pressure of just 10^{-12} that of Earth's at sea level. The outer reaches of this becomes ionized, also by solar radiation and energetic particles from Jupiter's magnetosphere, to form a very tenuous ionosphere. Some of the oxygen produced by the irradiation of the moon's surface is not thrown up into the atmosphere but remains in the surface ice, where it may eventually make its way down into the subsurface ocean. Some estimates suggest that the Europan ocean may have oxygen concentrations equal to those of the deep oceans of Earth. This raises an interesting question. Could the Europan ocean support some form of *life*?

Is There Life on (or, Rather, in) Europa?

Once, this question would have been greeted with laughter. But not any longer. The apparent presence of a subsurface ocean has changed the way we look at this moon from a biological perspective. Moreover, the discovery of ecosystems in the stygian depths of Earth's oceans demonstrates that life can and does exist in deep ocean environments completely cut off from sunlight. As we saw in the previous chapter, serious thought is now being given to the existence of simple forms of life in places as far removed from our familiar environment as the dwarf planet Ceres. Now, Ceres may have subsurface liquid water; Europa almost certainly does, so it is not surprising that this Jovian moon has rocketed to near the top of the list of possible life environments.

At first sight, it must be said that some of the abovementioned speculations as to the nature of this ocean do not make it seem very attractive. An ocean of brine might not be too bad so long as the brine solution is not too salty, but one of sulfuric acid sounds horrendous! Yet, simple organisms are known to thrive in similar environments right here on Earth, so we certainly cannot rule out their existence on Europa. On the other hand, we must not become too carried away with these prospects. Europan fish are unlikely to appear on any future spaceage menu. Single-celled organisms are probably the most that can be expected there, but finding even these would be an exciting discovery.

Discovering life on Europa will, nevertheless, not be an easy task. Unless some unmistakable sign of its presence turns up in the analysis of material erupted onto the surface from the ocean, getting to it will involve a sophisticated probe with the ability to bore down through miles of ice before releasing a small robotic submarine fitted with a battery of instruments for analyzing whatever it finds suspended in the water. This probe will need to be carefully sterilized to prevent any contamination from Earth. It will also need to withstand both the high radiation environment of the Jovian magnetosphere and whatever conditions it might encounter in the Europan ocean (acid? Strong brine?). Still, if continuing study of Europa confirms our present model of this moon and its ocean, a probe of this type may one day become a reality.

The suggestion of possible "lakes" enclosed within the moon's icy crust—cut off from both the deeper ocean and the surface—may also offer possible environments for life. Lakes of this type should also be more accessible, requiring less ice-boring to reach. At the moment, these are only hypothetical, but future research employing radar or even smaller and less sophisticated probes should prove or disprove their presence. Time, and continuing research, should eventually provide an answer to the intriguing question of Europan life.

Ganymede and Callisto

The two outermost of the Galilean moons are in many ways less "exotic" than the inner two and present surface features more in keeping with what is expected for Solar System objects largely devoid of atmosphere.

Ganymede is most notable for being the largest of the Solar System's moons. Having over twice the mass of our own Moon and a diameter of 3,273 miles (5,268 km) it would be classed as a planet if it orbited the Sun directly. Indeed, its diameter is close to 400 km *greater* than that of Mercury, the smallest of the major planets. It orbits Jupiter at a distance of 669,000 miles (1,070,400 km), completing one revolution of the giant planet in 7.155 days.

Physically, Ganymede is composed of equal amounts of rock and water ice, yet is a fully differentiated body with an outer ice mantle, a deeper silicate mantle and a small, molten, iron-rich core in which convection drives a magnetosphere, making Ganymede the only known moon to possess a true global magnetic field. There is also a very thin atmosphere with a particle density of about 7×10^8 particles per cubic centimeter which probably grades into an ionosphere at higher altitudes. This hypothetical ionosphere has not, however, been detected at the present time. Despite the extremely tenuous nature of the atmosphere, airglow has been detected, spread across the face of the moon, as well as bright spots at high latitude which appear to be aurora, probably triggered through interaction with Jupiter's magnetosphere.

It is thought that an ocean of saltwater may exist some 125 miles (200 km) beneath the icy surface. Needless to say, this ocean (if it really *is* there) is even less accessible than Europa's but might also rate as a possible site of simple living organisms. If the Europan ocean is eventually found to harbor life of some sort, the prospects of something similar lying deep within Ganymede will look brighter, but exploring this possibility will be a Herculean task unlikely to be attempted in the foreseeable future!

In contrast with the colorful volcanic patchwork of Io and the smooth cracked-egg profile of Europa, Ganymede sports many impact craters. Indeed, about one third of the moon's surface is saturated with them. This cratered terrain is thought to have survived almost since the formation of the moon some four billion years ago and to have preserved its ancient impact history. The heavily cratered regions are also the darkest areas on the moon's surface. The dark material covering these regions is believed to consist mainly of clays and organic compounds and probably samples the primordial material from which the regular moons of Jupiter accreted in the extreme youth of the Solar System. These regions of Ganymede may therefore act as a sort of time capsule preserving some of the earliest material from the days of planetary formation.

The greater part of the moon's surface appears lighter and less cratered but is crossed by extensive groves and ridges indicative of ancient tectonic activity. This may have been driven, as tectonic activity on Earth is driven, by internal heating from the decay of radioactive elements. Unlike the inner Galilean moons, radiogenic heat probably played a greater role in tectonic activity than heat from tidal stress unless Ganymede's orbit was more eccentric in the moon's youth. This, it should be noted, *is* a distinct possibility. In any case, whether radiogenic or tidal heating contributed the lion's share of the moon's tectonic activity, it seems that all of this happened a very long time ago. The "tectonic" terrain is very ancient; certainly younger than the darker heavily-cratered regions, though not greatly so according to all estimates.

Beyond Ganymede in the Jovian system, lies the outermost of the Galilean moons: Callisto. Orbiting the giant planet at 1,175,000 miles (1,880,000 km) this moon completes one full circuit of Jupiter every 16.69 days. It has a diameter of 3,013 miles (4,821 km) and an equal rock/ice composition similar to that of Ganymede. Unlike Ganymede however, Callisto is not fully differentiated. Although there is *some* degree of differentiation, there is no molten iron core. With increasing depth, the moon apparently becomes increasingly rocky and decreasingly icy and there might even be a small silicate core less than about 370 miles (600 km) in diameter.

The surface of this moon is noteworthy for being the most heavily cratered in the Solar System. It is literally saturated with craters. On a large scale, one part looks pretty much like any other area (craters, craters and still more craters!) but a closer and more detailed inspection finds that higher regions are covered by light frost deposits whereas the lower areas are dark. This is thought to be the result of sublimation of the more volatile constituents of the landforms, gradually degrading the features as lower areas are partially evaporated over time and higher ones coated by the re-freezing of some of the volatiles. What does not re-freeze escapes into the moon's atmosphere. Like Ganymede, Callisto has an extremely tenuous atmosphere which appears to consist mostly of carbon dioxide with, perhaps, a small amount of molecular oxygen. Having a particle density of just 4×10^8 particles per cubic centimeter and a surface pressure only 7.4×10^{-17} times that at sea level on Earth, the carbon dioxide atmosphere would be lost in approximately 4 h were it not being constantly replenished by sublimation from the moon's surface. The upper reaches of this atmosphere shade off into a rather intense ionosphere.

In common with Ganymede, the dark material on the surface of Callisto is probably representative of the stuff from which the moons were formed. It is thought to be close to that composing the carbonaceous D-type asteroids of the outer belt and in the system of Jupiter Trojans. A curious fact about the surface coloration is that the moon's hemisphere facing the direction of orbital motion is darker than the trailing one. It is thought that the trailing hemisphere has more carbon dioxide coating the surface and the leading one more sulfur dioxide, but just why there should be such an asymmetry remains a mystery.

The surface layer rests on top of an icy lithosphere 50–94 miles (80–150 km) thick. Magnetic observations suggesting that the Jovian field penetrates the moon are taken to indicate that a conducting layer at least 6 miles (10 km) thick exists below the lithosphere and the most likely candidate for this is, once again, an ocean of salt water. Accordingly, an ocean of between 30 and 120 miles (50–200 km) deep has been postulated to exist at depths of over 60 miles (100 km) beneath the surface. If this ocean does exist, the question of simple forms of life once more raises its head although, as is true of Ganymede, verifying or falsifying their presence will be no easy matter. All we can say at this moment in time is that should elementary life be found on (in!) Europa or some similar subsurface aquatic location, the probability of it existing deep within both Ganymede and Callisto increases. If, on the other hand, it is disproven on Europa (a far more difficult task, by the way—proving a negative is not easy!) the chances that it exists on these other moons looks slim. Incidentally, there may be other potential locations of subsurface life in the Solar System that are easier to examine than Europa and we will look at one of these in the next section. The balance of probability of life in Jupiter's system may come from discoveries in Saturn's, but more about this below.

The contrast between the innermost Galilean moon and the outermost could hardly be starker. While Io is the most volcanically active object ever seen, Callisto is—from the point of view of tectonic activity—as dead as Julius Caesar. Nothing happens on Callisto, and it seems that nothing ever has. The ancient surface gives no indication that tectonic or volcanic activity has ever taken place there. No mountain chains or ancient volcanic cones break through the fields of impact craters. Radiogenic heating has been insufficient to drive such activity and the tidal heating which played so vital a role in the evolution of the inner Galilean moons, and possibly even of Ganymede to some degree, is absent. In these respects therefore, Callisto is the least weird and least interesting of the Galilean moons. Yet, because of its distance from Jupiter it also avoids the bombardment of high energy particles that irradiate the inner members of Jupiter's family; the radiation levels at the surface of Callisto are low, making it a far more human-friendly location than the other Galilean moons. If ever a human outpost is established in the Jupiter system, it will almost certainly be situated on Callisto. There is much to be said for the quieter environment of this less-than-weird world!

Satellites of the Ringed Planet

Beyond Jupiter lies the spectacular Saturn, no longer alone in its ring system, but still unique within the Sun's family for the magnificence of this, its most characteristic feature. But Saturn is also a planet of many moons. At present, 62 have been discovered if we neglect the many moonlets within the ring system itself,; about 150 of which have been noted, undoubtedly just a sample of the total. This almost equals the number of discovered satellites of Jupiter but, because Saturn is so much further from Earth and its moons consequently fainter than those of equal intrinsic magnitude orbiting Jupiter, we can be pretty sure that others lurk undiscovered within its sphere of gravitational influence. Of the known moons of Saturn, 24 are regular and have prograde or direct orbits with inclinations close to that of Saturn's equatorial plane. The remainder consists of captured bodies similar to the irregular moons of Jupiter. These tend to be reddish in color, not unlike carbonaceous asteroids of the P and D type.

The Saturnian system is very top heavy in the way its mass is distributed. Some 96 % of the mass of the entire satellite system is concentrated in just one large moon; Titan. This object is second only to Jupiter's Ganymede amongst the Solar System's moons, having a diameter of 3,219 miles (5,150 km). By contrast, only 13 of the planet's moons have diameters in excess of 31 miles (50 km) with the smallest being just 2.5 miles (4 km) across. The largest of Saturn's moons, after Titan, are; Rhea (954.4 miles or 1,527 km), Iapetus (919 miles or 1,470 km), Dione (702 miles or 1,470 km), Tethys (664 miles or 1,062 km), Enceladus (315 miles or 504 km), Mimas (247.5 miles or 396 km), Hyperion (169 miles or 270 km) and Phoebe (133 miles or 213 km). All the rest have diameters in double or single figures. Of the large moons, all are regular with the single exception of Phoebe.

Some of the regular moons have noteworthy features. For instance, four of them are Trojans of other moons, that is to say, they inhabit the leading and trailing Lagrange points of the larger moons just as the system of Jupiter Trojan asteroids are found near the corresponding points of Jupiter's orbit. The large moon Tethys (located 184,137 miles or 294,619 km from Saturn and orbiting it every 1.89 days) is accompanied by the small Trojan moons Telesto at the leading and Calypso at the trailing Lagrange point and Dione (at 235,625 miles or 377,000 km from the planet and orbiting it every 2.74 days) is similarly accompanied by the leading Trojan Helene and the trailing Polydeuces. Trojan moons have been found only in the Saturnian satellite system.

Another odd feature of the system is a pair of co-orbital moons, or satellites that share (almost) the same orbit. The two moons are named Janus and Epimetheus and they each orbit Saturn at distances of around 94,600 miles (151,400 km) with orbital periods of just under 0.7 days. The orbits of these two moons are so close that they would collide if they attempted to pass each other. But instead of colliding, the gravitational action of the two causes them to swap orbits every 4 years. Now that is truly weird!

Yet other regular moons act as shepherds for the ring system (notably the moons Atlas, Prometheus and Pandora) and two (Pan and Daphnis) are found within ring gaps.

The system has some other odd beasts as well. For instance, the second largest of Saturn's moons—Rhea—appears to be absorbing plasma from Saturn's magnetosphere! That is the conclusion following the puzzling *Cassini* finding in 2005 of a depletion of energetic electrons in Saturn's plasma in the wake of this moon. One suggestion as to how it might be accomplishing this feat involved the presence of a faint dust ring around the moon. If this could be verified, it would make Rhea the only known moon in the Solar System to sport its own ring system. Nevertheless, while it might seem fitting that a moon of the most spectacularly ringed planet known to humanity should possess a ring of its own, subsequent searches have failed to find any trace of it. Curiously though, a very faint ribbon of material has been detected on the moon's *surface* girding its equator and it has been suggested that this may originate from material falling out of a ring system. Be that as it may, if no existing ring is found, the plasma observations remain a mystery. This moon is also noteworthy in possessing what might be one of the youngest impact craters on the inner moons of Saturn: a 30 mile (48 km) feature known officially as *Inkomi* and popularly as "The Splat" because of its prominent butterfly-shaped ejecta blanket. The moon itself is, as previously noted, second only in diameter to giant Titan amongst Saturn's satellites and orbits 329,443 miles (527,108 km) from the planet, taking just 4.52 days to complete a single revolution.

Speaking of impacts, it is clear from the scars they display that some of Saturn's moons have sustained massive ones in the past. Mimas, for example, carries a scar (the crater *Herschel*) on its leading hemisphere as large as a third of the moon's diameter! Tethys likewise must have been struck by something large as evidenced by a crater some 250 miles (400 km) wide (*Odysseus*) on its leading hemisphere and a canyon known as *Ithaca Chasma*, concentric with this crater and extending 270° around the moon.

Cold Volcanoes and a Hint of Life?

Two of the moons give evidence of tectonic activity and active cryovolcanism has indeed been observed on one of these. The active moon is Enceladus. At just 337.5 miles (504 km) it has the honor of being the smallest known geologically active object in the Solar System. It orbits the primary at 148,750 miles (238,000 km) every 1.4 days.

The other possibly active moon is Dione. Although no activity per se has been witnessed there, the surface appears to have been at least partially shaped in the past by tectonic forces and *Cassini* measurements indicate that, in common with Enceladus, this moon is shedding particles that subsequently become ionized and contribute to the plasma of Saturn's magnetosphere. Cryovolcanism is a possible source of these, though if it does exist on Dione, it must be at a much lower level than that experienced by Enceladus.

This latter moon is a very strange and interesting one. It is the only member of the Saturn system on which cryovolcanic activity has actually been observed and it is the smallest known body on which geological activity is known to occur. The "active" region appears to be concentrated near the moon's south pole, in an area found by the *Cassini* probe to be a lot warmer than expected and to be crossed by a system of odd-looking fractures graphically named "tiger stripes" by the *Cassini* team. These features run for about 81 miles (130 km) and it is from some of these that jets of water vapor and dust have been seen erupting and merging into a large plume off the moon's south pole. Dust from this plume replenishes Saturn's E ring. Gas molecules released in the plume are broken down and ionized, becoming the main source of plasma in Saturn's magnetosphere. Every second, more than 100 kg of material is erupted from the jets of Enceladus, continually contributing to these two features of Saturn.

The rest of this strange little moon's surface is not active in this way. Enceladus is not a lower temperature version of Io. Some of the surface is ancient and heavily cratered. Other parts are obviously a lot younger, being quite smooth and sporting only a few impact craters. Many of the younger plain areas are fractured, presumably by tectonic activity at some time. The moon is noted as being abnormally bright for its small size and is, in fact, one of the most reflective bodies in the Solar System, thanks to deposits of clean water-ice on its surface.

The activity of Enceladus is believed to come from the moon's 2:1 resonance with Dione. The internal heating caused by the resulting tidal flexing has probably resulted in an internal ocean of liquid water not unlike that thought to exist beneath the surface of Europa, albeit on a more modest scale. This ocean is now believed by a number of scientists to be the source of the erupting jets. Earlier, many thought that the jets originated in bodies of water closer to the surface (subsurface "lakes" as they might be termed), but more recently the discovery of salt in the jets has been taken as evidence for the water's origin deep within the moon where it is far more likely to have been in contact with rocky minerals and, in this way, to have acquired its saltiness.

Just as the Europan ocean has been cited as a possible home for simple forms of life, so its proposed counterpart within Enceladus has also been put forward as a possible life abode. If simple organisms do exist there, they are potentially easier to find than their hypothetical counterparts on Europa. It is an exciting thought that some of the "dust" particles erupted from this moon may actually be microbial spore! By contrast with the ice-boring exercise envisioned for a future life-search on Europa, a suitable probe

Fig. 4.4 Diagram demonstrating the electric circuit between Saturn and its moon Enceladus (Credit: NASA/JPL/JHUAPL/University of Colarado/Central Arizona College/SSI)

flown through one of Enceladus' plumes may be all that it needs to collect living (well, maybe not *living*!) organisms from another world. Even if a plume-sampling flyby did not find life, it would be an interesting exercise and one which could quite readily be incorporated with a raft of observations of Saturn and its moons.

One more interesting fact about Enceladus. *Cassini* discovered that it leaves a faint "footprint"—a dim auroral patch of light discernible at ultraviolet wavelengths—near the north pole of Saturn not unlike a fainter version of the Io footprint on Jupiter. The reason is similar; an electric circuit connecting the moon and its primary planet. Electrons are accelerated along lines of magnetic field and channeled into the planet's atmosphere causing the dim patch of light. The Enceladus footprint is, however, fainter than the "normal" Saturnian auroras. Nothing similar has, to date, been found at Saturn's south pole (Fig. 4.4).

The Crazy "Sponge" Moon

While it may not be as exciting from a potential astrobiological point of view as Enceladus, the craziest moon in the Saturn system—and arguably in the entire Solar System—is surely Hyperion. Orbiting the ringed planet once every 21.28 days at a distance of 925,625 miles (1,481,000 km) this moon is locked in a 4:3 mean motion resonance with the giant Titan. In other words, for every four orbits of Saturn made by Titan, Hyperion makes exactly three. It also has a very irregular and somewhat elongated shape, but neither its Titanic resonance nor its shape stands out as its oddest feature. The moon's tan colored surface has been likened to a sponge and that is not a bad description of this body as a whole. Hyperion is exceedingly porous. Its density is little more than half that of water and if splashed down in an ocean, the moon would bob around on the surface like a great lump of pumice stone! For it to have such a low density this body must be at least 40 % porous; nearly half of Hyperion is empty space! The surface appears to be quite old, judging from the number of impact craters found there. Craters having diameters in the 1.25–6.25 mile (2–10 km) range seem to be very common.

The moon's porosity is not, however, its only odd feature. Depending upon one's point of view, it might not even be considered the oddest one. Scientists who are more interested in dynamics than in composition might find the way it rotates even stranger than its low density. This moon is unique in the Solar System, as far as we are aware, in having chaotic rotation. That does not mean that it is tumbling all over the place as we watch, but it does mean that there are no well defined poles or equator. On short time-scales, it approximately rotates about the long axis at a rate of around 72°–75° per day, but on an extended time-scale, its axis of rotation wanders all over the sky and the long term rotation of this body is essentially unpredictable. Hyperion is truly in a class of its own amongst known Solar System bodies!

A Two-Faced Moon and a Tamed Centaur

A further two moons of Saturn deserve special mention (actually three deserve it, but one is so important that the entire following chapter has been reserved for it!). The objects concerned are Iapetus and Phoebe.

The first of these holds the distinction of being the most distant of Saturn's regular moons, orbiting at some 2,225,000 miles (3,560,000 km) from the planet and taking some 79.3 days to complete a single revolution. Its orbital inclination of 14.72° is also the highest of the planet's large satellites. But its chief claim to notoriety is the remarkable two-toned nature of its surface. Planetary astronomers have long been aware of the remarkable variations in the moon's brightness as it pursues its orbit around Saturn and deduced from this that one hemisphere must be markedly less reflective than the other. Closer views from space probes have dramatically confirmed this by revealing a striking contrast between a snow-white trailing and a pitch-black leading hemisphere. *Cassini* images reveal the dark material to spread from a large region (now known as *Cassini Regio*) close to the equator on the moon's leading hemisphere and extending both north and south to latitudes of around 40°. Both poles are bright, just like the trailing hemisphere.

Project 3: The Light and Dark Hemispheres of Iapetus

Iapetus is, of course, far too small and remote for its two hemispheres to be literally seen through Earth-based telescopes, but their presence can be ascertained by monitoring the changes in brightness of the moon as it pursues its orbit around Saturn. At its brightest (which happens when the moon is west of Saturn) it has been estimated as shining at magnitude 10.2 but fades to just 11.9, or nearly five times fainter, at its least luminous. A brightness change this large is readily apparent, and by comparing the moon's magnitude with that of stars of known brightness, the rise and fall in the luminosity of Iapetus can be charted over time as the moon pursues its orbit around its giant primary.

Apart from the remarkable color difference, there is little to distinguish the leading and trailing hemispheres of this moon. The entire surface appears ancient, heavily pockmarked by impact craters and devoid of any obvious signs of volcanic activity, past or present. The only really odd geological feature is an enormous 12.5 mile (20 km) high equatorial ridge spanning much of the circumference of the moon. The presence of this ridge gives the body a strange appearance which has been likened to that of a walnut, a little reminiscent of the "badly cracked egg" description of Europa! This ridge contains some of the loftiest mountains in the Solar System, reaching to over twice the height of Mt. Everest. But even this comparison, as it stands, scarcely does justice to the extreme elevation of this feature. The Earth is nearly 8,000 miles in diameter; Iapetus just 919. Mt. Everest is a little under 0.0007 times the Earth's diameter. But the altitude of the Iapetian equatorial ridge is a full 0.014 times that moon's diameter. Proportionally speaking, Everest would need to be around 109 miles high to equal the equatorial ridge!

The reason why Iapetus has such a feature is not understood. Some researchers suggest that the ridge might be a remnant of the oblate shape of a rapidly-rotating youthful Iapetus. This hypothesis suggests that early in its life, heating by short-lived radioactive isotopes such as aluminium-26 caused this ice/rock moon to be in a more plastic state. If it was both plastic and very rapidly rotating, the moon may have deformed into the famous walnut shape before decay of the heat source and consequent cooling "set" it into that form.

Another hypothesis sees the ridge as due to the upwelling of icy material from beneath the surface while yet another postulates the collapse of an ancient ring around the moon, depositing material along the equatorial zone. A variation of the latter postulates that this moon once had a moon of its own; a satellite of a satellite which, over time, spiraled in toward Iapetus before eventually disrupting and crashing to form the great ridge.

The ring hypothesis is probably the least likely explanation, as material from a collapsing ring appears inadequate to build a ridge of the required elevation. The "second-order moon" model may be in a stronger position, but problems nevertheless remain with any model that looks toward some external factor to explain the ridge.

Evidence of tectonic faulting through the ridge (not a common sight on Iapetus), suggests that the formation of this feature was more likely driven by internal rather than external forces.

Iapetus has another claim to fame. It has some of the biggest landslides known. Evidence of slides of the type known as "long-runout landslides" or "sturzstroms" some 50 miles (80 km) long has been found along the crater walls and other slopes of this moon and their height/length ratio initially presented planetary scientists with a puzzle. The landslides of Iapetus sometimes travelled as far as 30 times their height. By comparison, landslides and avalanches on Earth seldom make it more than twice their height. Ice was suspected as having some involvement in the length of these slides, however experiments with icy mixtures on Earth still failed to match the scale of these events. Somehow, the ice on Iapetus acted as though it possessed a lower friction coefficient than the familiar ice of Earth and therefore ice/rock mixtures on this moon could travel to greater distances before being brought to a stop. Is there something weird about the ice of Iapetus? Probably not. The most likely hypothesis involves ice melt thought friction. Friction between the sliding material and the surface over which it is moving generates heat. If the sliding material is comprised of both rock and ice and if the frictional heat is sufficient, pieces of ice will melt and cause the debris flow to become more fluid and travel further than one composed entirely of solid particles.

But back to the strange color dichotomy of Iapetus' surface. Why does this moon possess such a unique feature? The answer almost certainly involves the other distant moon which deserves our special attention; Phoebe.

Phoebe is unusual in several respects. Unlike the other satellites discussed here, it is an *irregular* moon, that is to say, it did not form from an early disk of material as the regular Saturnian satellites are believed to have done, but was initially a freely orbiting member of the Sun's family that became ensnared in Saturn's gravitational net and captured by the giant planet. It is far and away the largest of Saturn's irregular moons and travels around the planet in a retrograde orbit some 8,043,563 miles (12,869,700 km) distant, taking 545 days to make a single revolution. It rotates about its axis every 9.3 h, is almost spherical in shape, quite heavily cratered and has a relatively high density of 1.6 times that of water, indicative of a rock and ice composition. The surface of the moon is very dark and is covered by a mixture of water ice, frozen carbon dioxide, phyllosilicates and organic compounds. There is a possibility that iron-bearing minerals also add to the mix.

When the first long-distance space probe images of Phoebe were received, one astronomical writer commented that what was being observed was probably "the encrusted nucleus of a giant comet". That seems not to have been too far from the truth, as most astronomers believe that prior to its capture by Saturn, this moon was a centaur or a Kuiper belt object. The close-up images of this object provided by *Cassini* may have provided us with the best likeness of Chiron and its siblings that we are likely to see in the foreseeable future!

Phoebe is now known to be associated with a faint ring; in terms of diameter, the largest in Saturn's system of rings and one which is composed of dark particulate matter believed to have been ejected from Pheobe's surface by meteoroid impacts.

It is this free-floating material from the moon's dark surface that solves the mystery of the two-faced Iapetus. Like Phoebe itself, the particles thrown up from its surface orbit Saturn in a retrograde direction, gradually spiraling inward toward the planet. As Iapetus pursues its (direct) orbit around the planet, it sweeps up the particles with its leading hemisphere. However the full story of Iapetus' dark hemisphere is not simply that it has been covered by Phoebian debris. The situation is a little more complex than that. Being dark, this material absorbs sunlight more readily than lighter-colored materials, converting it into heat and thereby raising the surface temperature of the regions blanketed by Phoebe's detritus. This in turn leads to a thermal runaway process whereby ice evaporates; exposing darker material underlying the icy surface covering which (in its turn) absorbs more sunlight and so on. The upshot is that ice evaporates from the leading hemisphere and re-condenses on the trailing, increasing the reflectivity of the latter as it darkens the former and creating the extreme contrast in reflectivity and color that we observe today.

Although it is generally thought that meteoroid impacts alone are responsible for kicking up dark debris from Phoebe, the present writer wonders if another factor might also operate, at least occasionally. Phoebe, as already stated, is believed to be a captured centaur or Kuiper belt object and as such is basically "cometary" in nature. Chiron and some other centaurs have been observed to display cometary activity and although no such thing has been witnessed in association with Phoebe that does not necessarily mean that the moon has *never* displayed this sort of activity. The impact of a large meteoroid might be sufficient to trigger a bout of activity on the moon and there may be other potential triggers as well. If outbursts of activity do occur from time to time, these presumably could throw significant quantities of debris into the space surrounding Phoebe; material capable of contributing to the Phoebe ring and to the dark particles that came to rest on the leading hemisphere of Iapetus. Although witnessing a cometary outburst on Phoebe is highly unlikely, any suspicion that such a thing might be happening should be swiftly (albeit cautiously!) reported.

Moons of the Ice Giants

Orbiting far beyond the gas giants, the two so-called "ice giant" planets, Uranus and Neptune, are also well endowed with moons, although their known numbers are more modest than those of either Jupiter or Saturn. Uranus, the nearer of the ice giants, has 27 known moons and Neptune, 13. Nevertheless, because these planets and their satellites are so far from Earth, it is highly likely that other faint members of their satellite systems still await discovery.

Companions of Uranus

Turning first to the Uranian system, its members are normally classified into three groups; the inner moons (13 in number), major moons (5 of these) and 9 irregular moons or objects that were originally orbiting the Sun but which at some time in the remote past were captured by the planet's gravitational field. These latter move in highly inclined orbits, all but one being retrograde, at large distances from the planet. The most distant yet discovered is Ferdinand, a small moon orbiting the planet at a remarkable distance of 13,063,125 miles (20,901,000 km)! The only one of the irregular moons known to be moving in a prograde or direct orbit is Margaret, orbiting Uranus at an average distance of nearly nine million miles (14,345,000 km) in an orbit that currently exceeds all other Solar System moons in terms of eccentricity (0.7979). (Its mean eccentricity, however, still falls below that of Neptune's Nereid. Like all the irregular moons of Uranus, its orbit is significantly perturbed by the distant Sun.) Margaret takes about 4.6 Earth years to complete a single orbit of the planet.

By contrast, the inner moons of Uranus—all of them regular satellites believed to have formed together with the planet—hug their primary at distances ranging from 31,094 miles (49,751 km) for Cordelia to 61,084 miles (97,734 km) in the case of Mab. The largest of this group of moons is Puck at just over 101 miles (162 km) diameter and the smallest is the tiny Cupid with a diameter of just 11.3 miles (18 km). All are dark objects apparently composed of a mixture of water ice and dark, probably organic, material and are intimately associated with the Uranian ring system. Indeed, it is thought that these rings probably represent the remains of one or more former moons that were destroyed through mutual collision or by the impacts of incoming bodies. The outermost of the inner moons, Mab, seems to be the source of the so-called Mu ring of Uranus. Closer in, the moons Cordelia and Ophelia act as shepherds to the Epsilon ring. All of these inner moons whip around the planet in less than a day and computer simulations reveal that they are in danger of perturbing each other into intersecting orbits that might eventually result in mutual collisions. Perhaps that has already happened to now-vanished inner moons; the main system of rings being their memorials. Projecting into the future, it seems that one of these inner moons—Desdemona—may collide with either the moons Cressida or Juliet at some time within the next 100 million years.

Contrasting with these small moons, the five major satellites of Uranus are significant bodies which would be classed as dwarf planets if they were found orbiting the Sun. In other words, they have achieved hydrostatic equilibrium and are approximately spherical in shape. The largest of the large moons—Titania—is approximately 986 miles (1,577 km) in diameter and is around 20 times less massive than Earth's Moon. Having a diameter of just 295 miles (472 km), the innermost Miranda is the smallest of the quintet; the remaining being Oberon at 952 miles (1,523 km), Umbrial at 731 miles (1,170 km) and Ariel at 724 miles (1,158 km). In terms of distance from the primary, Miranda orbits at 80,869 miles (129,390 km) and has an orbital period of 1.4 days, Ariel 119,388 miles (191,020 km) every 2.5 days, Umbrial 166,438 miles (266,300 km) every 4.1 days, Titania at 272,444 miles (435,910 km) every 8.7 days and Oberon, the most distant of the five at 365,000 miles (583,520 km), orbiting the planet once every 13.5 days.

With the exception of Miranda, the major moons appear to be composed equally of ice and rock. The ice is predominantly water ice, although frozen ammonia and carbon dioxide may also be present to a lesser degree. Miranda seems to consist primarily of water ice with only a small rocky content. Some differentiation—into rocky cores and icy mantles—may have taken place within the two largest and it is even possible that liquid water "oceans" exist between the core and mantles of this pair. Once again, this raises the perennial question of simple life. And once again, proving or disproving the case will be a Herculean task considering the depth at which these oceans exist—assuming that they exist at all!

The five moons have surfaces that are heavily pockmarked with impact craters, but (with the exception of Umbrial) they also show signs of tectonic activity, mostly in the form of canyons, although oval features, described as looking like racetracks, have also been found on the surface of Miranda. Presumably these *coronae*, as they are officially known, are not the venues of Mirandian horse races. More likely, they are sites of past upwelling of warm ice from beneath the moon's surface crust.

Ariel is somewhat less heavily cratered than its neighbors, indicative of a surface not quite as ancient as those of the other four. Presumably this moon experienced some tectonic activity more recently than its siblings. The internal heating responsible for the tectonic activity of these moons is thought to have resulted from past orbital resonances between them. A 3:1 resonance between Miranda and Umbriel and a 4:1 between Ariel and Titania are believed to have been responsible for internally heating both Miranda and Ariel. These resonances are no longer present, but their former existence may explain a curious feature of Miranda's orbit; it has an inclination of 4.34°. Although this would be thought unexceptional in the wider scheme of things, it is actually pretty high for a moon orbiting so close to its primary. An early orbital resonance might solve this problem, as well as explaining the evidence of internal heating in such a small and icy object as Miranda.

All of these moons are thought to have formed either from a disk of material left over from the formation of the planet, and surrounding it like a miniature planetary accretion disk, or else from debris ejected by the giant impact that is hypothesized to have flipped Uranus onto its side. The moons, by the way, orbit the planet at low inclination, which means that the system of regular satellites shares a similar orientation with respect to the plane of the planetary system. The distant irregular moons, on the other hand, have a wide range of inclinations.

The sharing of Uranus' axial tilt by these major moons has a weird effect on how a hypothetical observer on the surface of one of their number, at the time of summer solstice, would view the Sun. In short the Sun would seem to move in a circle around the celestial pole, being at its closest about 7° away from it. Near the equator, the Sun would be seen nearly due north or south, depending upon whether it is southern or northern summer solstice, but at latitudes higher than 7°, it follows a circular path around 15° in diameter, remaining visible throughout the entire process. This makes the terrestrial "midnight Sun" look very tame by comparison!

The Strange Little World of Miranda

As we are principally interested in "weird" moons, the one of particular interest in the Uranian system has to be Miranda. Although not as "weird" as some of the other worlds in the Solar System, this little moon nevertheless has some truly odd features. For a start, it is one of the few known Solar System objects whose pole-to-pole circumference is smaller than the equatorial (more prolate than oblate, in other words). Like Saturn's Enceladus, it has also experienced a good deal of internal activity for so small a body. Although present activity has not been observed, there are signs of past upwellings of warm ice from its interior as well as possible eruptions of icy magma in cryovolcanic outbursts. But the real oddities are its systems of escarpments and high cliffs. Close images of the moon show a veritable jumble of blocks and troughs

Fig. 4.5 Miranda, showing the high scarp Verona Rupes (Credit: NASA)

which give the appearance of rift valleys or similar formations. However, the scale is tremendous. Scarps appear to rise to an altitude of at least 3 miles (5 km)! Considering that the moon is less than 300 miles in diameter, this is proportionally equivalent to 80-mile-high cliffs on Earth! It has been estimated that, given the height of its cliffs and weakness of the moon's gravity, if someone were to fall from the top of one of Miranda's cliffs, it would take them 15 min to reach the bottom! Plenty of time to contemplate one's fate we may suggest (Fig. 4.5).

The Children of Neptune

The second ice giant—Neptune—is known to have 13 moons, however this system is so far from Earth that it would not be at all surprising if numerous small satellites still await discovery. Of the known Neptunian moons, six are regular and six irregular or captured, plus one which seems to have a foot in each camp, as will be explained below. Unusually, the first known moon of this planet was an irregular one (Triton) and, equally unusual, it is by far the largest of all of Neptune's satellites. The first genuinely regular moon, Larissa, was not discovered until 1981.

The regular moons—Naiad, Thalassa, Despina, Galatea, Larissa and Proteus—all orbit the planet at relatively close distances, ranging from 30,142 miles (48,227 km) for Naiad to 73,529 miles (117,646 km) for Proteus. The former completes one revolution of the planet in just under 8 h, the latter in 1.2 days. Naiad and Thalassa orbit between two of Neptune's rings (the Galle and LeVerrier rings) while Despina lies just beyond the latter ring and is suspected of being its "shepherd". Galatea orbits a small distance within the very narrow but more prominent Adams ring and resonances between this moon and ring particles are thought to be at least partially responsible for the arcs displayed by this feature. None of these moons is very large; the biggest—Proteus—has an average diameter of 263 miles (420 km) while the smallest—Naiad—measures just over 41 miles (66 km) across. Larissa, with an average diameter of around 125 miles (200 km) is somewhat elongated while Proteus has a very odd polyhedron shape with a number of flat or even slightly concave facets some 94–156 miles (150–250 km) across. This peculiar form probably reflects a history of impact-generated disruption. The surface of this moon also appears quite battered, being heavily crated as well as showing a number of linear features that probably tell the story of a violent past. The largest of this moon's craters (Pharos) is at least 94 miles (150 km) across; quite a scar for so small a moon!

All of the regular inner moons of Neptune are rather dark bodies, thought to be composed of water ice mixed with dark and probably organic material. In terms of composition, they closely resemble the inner moons of Uranus and most probably formed from an accretion disc of material remaining after the primary had coalesced. Typical of regular moons, their orbits lie close to the equatorial plane of Neptune itself.

The irregular moons present a far less compact and orderly system. The innermost of these—Triton—orbits at an average distance of 221,724 miles (354,759 km) and has an orbital period of 5.9 days. Its period of rotation is synchronous with its orbital period of Neptune, so that it always shows the same face to the planet. The outermost Neptunian moon thus far discovered—Neso—has an average orbital diameter of 30,803,125 miles (49,285,000 km) and takes just under 27 *years* to complete a single orbit of the planet! This is the furthest flung moon of any planet yet discovered. Four of the irregular moons (including Triton, the largest) have retrograde orbits.

Included amongst the retrograde satellites is an object that is believed to have once been an "indigenous" moon of the planet. This is the object mentioned earlier as having a foot in both camps as it were. With a diameter of 213 miles (340 km) it also rates as Neptune's third largest moon and is the second to have been discovered; Nereid. This object currently moves in an orbit with the highest average eccentricity of any known moon in the Solar System (0.7507) and has an average distance from the planet of 3,446,136 miles (5,523,818 km) and an orbital period of about 1 year. It's strange orbit and split personality between regular and irregular is thought to have been caused by the capture of Triton. This big moon so disrupted Nereid's orbit that it almost hurled it out of the Neptune system altogether; a fate which probably befell a number of the planet's original satellites thanks to the bullying tactics of this large intruder. Others may have been driven into collisions with one another, leaving their remains in the system of rings surrounding the planet.

Triton: The Giant Interloper

Triton is indeed a strange beast. Having a diameter of 1,691 miles (2,705 km) it is far and away the largest of Neptune's satellites and actually contains most of the mass of the entire satellite system. Yet, in contrast to the largest members of all other known satellite families, Triton is not native to Neptune at all! It is the guest that came to stay; an interloper in the household. Triton is thought to have once orbited the Sun as a Kuiper belt object before being captured by the ice giant. Like three of the other irregular moons of Neptune, Triton's orbit is retrograde, making it the largest of the Solar System moons to orbit its primary in this way (Fig. 4.6).

Interestingly, an early hypothesis suggested that Triton and Pluto were both originally moons of Neptune. At some time in the distant past, so this line of reasoning went, the two objects dynamically evolved in such a way as to bring them into close contact,

Moons Galore! 231

FIG. 4.6 Triton as imaged by *Voyager 2* (Credit: NASA)

resulting in such severe orbital perturbations as to expel Pluto from the system altogether and effectively turn the orbit of Triton completely around into a retrograde path. This hypothesis looked quite reasonable on paper—literally on paper, as a flat representation of the orbits of Neptune and Pluto can give the idea that these two bodies can approach very close to one another. Nevertheless, a three-dimensional model uncovers quite a different situation. Pluto cannot make a very close approach to Neptune because of the inclination of its orbit to the plane of the planets. Moreover, the discovery of other Pluto-like (and Triton-like!) bodies in the Kuiper belt during recent decades has turned this early idea completely on its head. Far from Pluto being an escaped Triton-like moon of Neptune, it now seems that Triton is a captured Pluto-like object from the Kuiper belt!

This body is one of only two moons in the Solar System possessing an atmosphere thick enough to generate what might loosely be called "weather". By the standards of Earth however, the atmosphere is perilously thin; having just one seventy thousandth of the pressure of Earth's air at sea level. Needless to say,

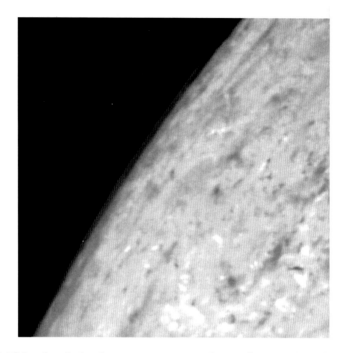

Fig. 4.7 Thin clouds in the tenuous atmosphere of Triton (Credit: NASA)

the "weather" that occurs on this moon is confined to a constant haze (believed to consist of hydrocarbons and nitrates), prevailing winds sufficient to cause "geyser" plumes to drift (more of this shortly) and some clouds of condensed nitrogen drifting at altitudes of around 0.6–2 miles (1–3 km) above the surface (Fig. 4.7).

The atmosphere itself seems to primarily consist of nitrogen with small amounts of methane and carbon monoxide. Similar to Pluto, nitrogen probably sublimates off the surface of the moon into the atmosphere.

Nitrogen ice covers the surface of Triton, together with amounts of frozen carbon dioxide and water ice. Beneath this, a crust of water ice—as hard as rock—overlays a deeper icy mantle and, deeper again, a core of rock and metal. This core is thought to comprise as much as two thirds of the moon's entire mass. It is possible that an ocean of liquid water exists deep within the moon, once more raising the issue of underground alien biology, even in this seemingly unlikely place!

Much of Triton's surface is relatively young, judging by the few impact craters that have been found on the moon. The majority of those that have been noted are confined to what appear to be older plains. Rifts and scarps cross the plains, indicative of geological activity and a good deal of the surface has apparently been rejuvenated by cryovolcanic eruptions of lava consisting of a mixture of water and ammonia ice. There is enough rock deep within the moon for the heat from radioactive decay to drive geological activity on this object, so the presence of cryovolcanism is not too surprising.

However, the weirdest feature of Triton is surely the field of erupting jets located between 50° and 57° south. These are often referred to as "geysers" and, in one sense that is a fair enough description. Nevertheless, they do not seem to be deep volcanic features like the erupting hot-water springs of Earth known by that name. Tritonian "geysers" are principally jets of nitrogen gas, laden with dust particles, that erupt through the icy surface to heights as great as 5 miles (8 km) and which can remain in a state or eruption for up to one Earth year. The plumes of dust released drift downwind and settle into streaks across the ice, marking the location of these eruptions. All of the "geysers" observed are located beneath the sub-solar point and, weak though the Sun's heat must be at Triton's distance, it is apparently strong enough to be the trigger for these remarkable features. The favored explanation runs as follows: There is probably a layer of translucent nitrogen ice on the surface of the moon which allows solar rays to penetrate and reach a darker subsurface layer of dusty nitrogen ice. Sunlight slowly heats the darker material until some of the ice sublimates into gaseous nitrogen. The more ice turns to gas, the greater will the pressure become on the surface layer until, finally, that layer cracks and pent-up gas jets forth in a tremendous torrent, blowing out fine dust particles in the process (Fig. 4.8).

Does this sound familiar to the reader?

Remember the carbon dioxide jets on Mars? The process is essentially the same, except that in the far colder environment of Triton, carbon dioxide is replaced by nitrogen!

One final point about Triton. It is living on borrowed time. Because it orbits Neptune in the "wrong" direction, tidal interaction with the planet is drawing it ever closer. (This, it will be recalled, is the opposite of Earth's Moon, which is slowly becoming more remote). In about 3.6 billion years time, it will reach

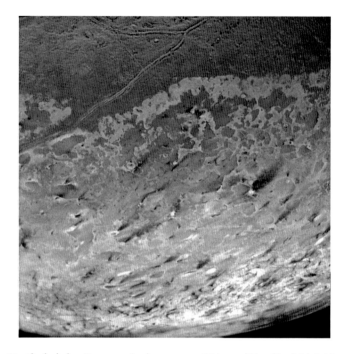

FIG. 4.8 Trails left by "geyser" plumes on Triton (Credit: NASA)

Neptune's Roche Limit; the distance at which tidal effects will tear the moon apart. Either large pieces of the moon will hurtle down into the planet's atmosphere or Neptune will have a bright new ring; memorial of its former strange moon!

Moons of the Minor Bodies

Moons are not confined to full-sized planets. Asteroid satellites are not uncommon and these range from little more than orbiting pieces of rock to bodies that are more correctly described as double asteroids or binary asteroids. Such is the asteroid 617 Patroclus; a Jupiter Trojan now known to consist of two bodies of closely-similar size orbiting a common barycenter or mutual center of gravity. The slightly smaller body has been given a satellite-type designation and name (Patroclus 1 Menoetius) but the similarity in size between Menoetius and Patroclus means that the former is not a moon in the true sense of that word (Fig. 4.9).

FIG. 4.9 An asteroid and its moon. Asteroid Ida and its tiny moon Dactyl as imaged by the Jupiter-bound *Galileo* spacecraft (Credit: NASA)

Moons are also known to accompany many of the icy denizens of the Kuiper belt at the fringes of the Solar System. Many of these are larger, compared with their modest primaries, than the moons of the major planets with the single exception of the Earth's oversized companion. Like Earth's Moon, the satellites of these smaller bodies are thought to have arisen through impacts with other objects—asteroids in the case of the denizens within the gulf separating Mars and Jupiter and other large icy bodies for the inhabitants of the Kuiper belt and more distant regions. Support for this thesis comes from the observation that moons are more common around the larger Kuiper belt bodies. Of the four brightest objects of this class, three are known to have moons while only about 10 % of the fainter observed ones within this region are similarly accompanied. If a body larger than about 625 miles (1,000 km) is struck by a large object, its gravitational pull will be sufficient to hold the ejected debris in orbit permitting the coalescence of this into an orbiting body or bodies. Smaller objects will have insufficient gravity to stop collisional debris from escaping into space, although large chunks of material may sometimes be captured in orbit and become moons.

There is little point in dealing with individual moons here. Instead, just one of these mini systems will be looked at; the one which is most widely known and, of the known examples, in many respects the strangest. The system is, not surprisingly, Pluto and its system of moons.

(As an example as to just how fast things can change in the field of astronomy, the writer was already at work on this section when the announcement came that a fifth moon had been added to Pluto's retinue! Who knows how many more might have been found by the time you read these words?).

Despite protests to the contrary, Pluto is now generally accepted as a dwarf planet. The system of moons under its governance may just as easily be seen as a dwarf system also. Pluto's is the most compact known satellite system and occupies just 3 % of what planetary astronomers call the Hill radius, i.e. the distance from the primary where bodies can be held in stable satellite orbits.

The largest of the Plutonian moons is Charon, first identified in 1978 by James Christy. It is a large body containing 11.6 % of Pluto's mass and having a diameter of 754 miles (1,207 km). It is located just 12,231 miles (19,570 km) from Pluto itself. So close are these two bodies and so massive is Charon with respect to Pluto that they effectively form a binary system. The barycenter of the system lies outside of either body, something that does not occur in any major planetary satellite system (the Earth/Moon system comes closest, but the barycenter still lies within Earth). Charon orbits the barycenter with a period of 6.4 days, essentially identical with the time it takes Pluto to turn once on its axis. This has the curious effect of causing each body to remain in the same position in the other's sky, or invisible at the same spot beneath the horizon depending upon the hypothetical observer's location. Just as the rotational axis of Pluto effectively lies on its side, the Pluto/Charon system (and, indeed, the Pluto satellite system as a whole) also lies on its side. In this sense, the system resembles a scaled-down version of the system of the regular, though not of the irregular, satellite system of Uranus.

We will look more closely at Charon shortly.

In the meantime, what of Pluto's other moons? These are very different from Charon in size and all orbit further out. The first two of these small moons were discovered using the *Hubble Space Telescope* on May 15, 2005. Now known as Nix and Hydra, they orbit the barycenter at distances of 30,438 miles (48,700 km) and 40,500 miles (64,800 km) respectively and, depending upon their exact degree of reflectivity, are estimated to have diameters somewhere between about 29 and 86 miles (46–136 km) for Nix and 37.5 and 105 miles (60–168 km) for Hydra. Curiously, their orbits lie close to (though not exactly coinciding with) the 1:4 resonance (Nix) and the 1:6 resonance (Hydra) with Charon. The actual orbital periods of these moons are 24.9 and 38.2 days respectively. From observations of their color, they appear to be similar in composition to the surface of Charon.

A fourth moon—for the present simply known as S/2011 P1—was announced in July 2011. This object is estimated at between 16 and 42 miles (26–68 km) in diameter and orbits the barycenter at about 37,000 miles (59,000 km), in other words, between the orbits of Nix and Hydra. This moon is also very nearly, although once again not exactly, in a resonance with Charon; in this instance, the 1:5 resonance (orbital period, 32.1 days). Then, on July 10, 2012, a fifth moon—announced as S/2012 (134340)1—(134340 being Pluto's designated number in the catalogue of minor planets) was found orbiting at around 26,000 miles (42,000 km) from the barycenter or between the orbits of Charon and Nix. This moon has a period of 20.2 days and lies very close to the 1:3 resonance with Charon. It is estimated as being just 6–17 miles (approximately 10–25 km) in diameter. The presence of these small moons has raised questions as to whether a ring of particles, thrown up from their surfaces by meteoroid impacts, encircles Pluto. Searches have not yet found such a ring which, if it exists at all, must therefore be very faint. The searches have not come up completely empty handed however, as both the 2011 and 2012 satellite discoveries were made during the hunt for possible rings of Pluto. Hopefully, *New Horizons* will give a definitive answer to this question of a ring when it cruises past the Pluto system in 2015.

The satellite system of Pluto might be interesting because of its compact nature and its strange orientation but the object that makes it truly exceptional is Charon. Like its primary, Charon is composed of a mixture of ice and rock, but astronomers are not sure if these exist in roughly equal proportions throughout the satellite or whether Charon is differentiated into a rocky core surrounded by an icy mantle. Either way, its density is less than that of Pluto (1.65 times that of water as against 2.03 times for Pluto), indicating that is contains proportionally more ice and less rock than its primary. The differentiated model received something of a boost in 2007 when the Gemini Observatory detected the signature of patches of crystals of water ice and ammonia hydrates on the moon's surface. The fact that the ice is in crystalline form strongly suggests that it has been deposited in at least relatively recent times, as even at the distance of Charon, the Sun's radiation is still sufficient to degrade ice to an amorphous state after roughly 30,000 years. In the absence of an atmosphere capable of supporting snow and hail, the most likely means of depositing fresh ice is through cryo-geysers. Fresh ice implies fresh—maybe currently active—cryo-geyser activity on the moon and this in turn adds weight to the differentiated model of Charon. Once again, we await further data from *New Horizons* when it arrives in the Pluto system in July 2015. It would be great if a cryo-geyser is caught in action.

Although the compositions of Pluto and Charon appear to be roughly similar, it is noteworthy that the surface of Charon is not coated with methane and nitrogen ice in the manner of the primary. It is this frost on Pluto that gives the dwarf planet its reddish color and which boils off near perihelion forming a temporary atmosphere before refreezing again as Pluto recedes from the Sun. Charon lacks this coating, accounting for the observed color difference between the two objects; Charon's grayish appearance contrasting with the reddish hue of Pluto. Incidentally, speaking of color, Charon has been found to be lighter or brighter near the equator and darker at the poles, with the southern polar region being somewhat darker than the northern. Presumably, helping to sort out the reason for this will be another task for *New Horizons* in 2015.

The moon's surface depletion of the very volatile ices of methane and nitrogen has been taken as supporting evidence for the hypothesis that this moon was formed from a ring of debris thrown up following an impact between Pluto and another large Kuiper belt object. This, as already noted, is a popular hypothesis for the formation of satellites of the Kuiper belt bodies. One would expect much of the store of very volatile ices to have been lost during the collision and in the subsequent process of accretion of the debris. Moreover, Charon's low rock content relative to Pluto has also been taken as support for the idea that Charon is comprised to a great extent of material thrown out from Pluto's mantle by the hypothetical collision.

More recent work suggests however, that the impact thesis needs modifying. The problem is, although the icy nature of Charon relative to Pluto should be a consequence of the hypothesis, further assessment indicated that the difference should be even *more* marked than it appears to be. In other words, Pluto should be even rockier and Charon even less rocky than they are. The newer "modified" impact hypothesis postulates that Charon *itself* was the impacting body, not the result of accretion of debris following a strike by a third object. According to this hypothesis, Charon and Pluto were initially unrelated Kuiper belt objects, but that at some time in the distant past Charon literally "bumped into" Pluto in a very low-velocity collision that was nevertheless sufficient to deplete the former of most of the very volatile ices that presumably coated its surface. Yet, the impact was insufficiently violent to shatter either body. Following this gentle impact, the two bodies assumed orbit around a common center of gravity and the Pluto/Charon pair was born. Such a scenario also appears to be in agreement with the previously noted discovery that larger Kuiper belt objects are more likely to have moons than smaller ones. The larger objects should be better at holding onto bodies that nudge into them at low relative velocity. Perhaps both scenarios are at work amongst Kuiper belt objects.

The other moons of Pluto were at first thought to have been fragments broken from Charon. They were supposed to have initially orbited closer to Pluto than they do today but later migrated outward as Charon assumed its present orbit, tugging them into the positions of near resonance where we now find them. On the other hand, more recent orbital investigations suggest that this process would place the moons in orbits of higher eccentricity than the near-circular ones that they in fact follow. Accordingly, a different scenario has been proposed in which the smaller Plutonian moons were independent objects, captured by Pluto, that have had their orbits reduced and made more nearly circular by Charon's influence, eventually being brought into near resonance with the large moon. This would be a way around the eccentricity problem, but it may run into another difficulty in so far as both Nix and Hydra are (as far as can be ascertained from observations of these faint bodies) very similar in color, and presumably therefore in surface composition, to Charon. This poses no problem if they are fragments of that moon, but it would be quite a coincidence if they originated as unrelated bodies. One solution might be to suppose that the surfaces of these moons are covered with material driven off Charon by impacts or even expelled through cryo-geysers. The bulk composition of the small satellites might then be very different from that of the large moon but, because we only see the surface anyway, we would be tricked into thinking that they looked like little pieces of Charon. Hopefully this will be yet another subject on which *New Horizons* will shed more light!

This short overview of the "weirdest" moons of our system has demonstrated that, if nothing else, many of these bodies are far from the uninteresting hunks of rock that they were assumed to be by so many for so long. Both in their relationship with their primaries and concerning their intrinsic properties, there are many things about these bodies that strike us as being both interesting and strange. Nevertheless, the reader will have noticed that one moon has been simply brushed over, a little like the proverbial elephant in the room that nobody wants to mention. There is a good reason for this. This very strange moon—we refer to Titan of course—deserves a chapter on its own. Arguably, it is the weirdest world in the Solar System and the one which comes closest to the bizarre imaginary planets of science fiction with their exotic landscapes, strange atmospheres and maybe—just maybe—totally alien forms of life! To this weird world, we shall now turn.

5. Titan: The Weirdest World in the Solar System?!

Science fiction written prior to the dawn of space exploration nearly always pictured other worlds as weird and wonderful places. Exotic landscapes, exotic life—exotic just-about-everything was the order of the day. Then came real space exploration and the romantics amongst us must have felt a tinge of disappointments at the rather bland scenes being transmitted back to Earth. Gone were the jagged mountains of the Moon, the canals and strange plants of Mars and global oceans or bubbling oil fields on Venus. True, the human landings on the Moon and the robotic ones on Mars needed to set down at the safest locations available, and these "safe" regions were mostly rather dull in appearance, but the impression that the worlds around us were mostly unexciting fields of broken rocks was difficult to avoid.

And then there is Titan! Although the first scenes from its surface in 2004 did not look too different from the early ones transmitted back from the surface of Mars in 1976, the orbiter of the *Cassini-Huygens* probe, and the progressive analysis of the data sent back since its arrival at the moon, revealed a world that is the closest yet found to those weird imaginary planets of science fiction. Here at last is a "never-never land"; a place of thick atmosphere, seas of liquid hydrocarbons, methane storms, liquid methane rivers carving valleys in landscapes of rock hard water ice and dunes of organic dust. All that seems to be missing is alien life and, if some speculations on this matter turn out to be correct, this appearance may be deceptive. In fact bold, though not wild, speculation raises the possibility that this moon may harbor forms of life whose metabolism is such that, were they to be introduced into our terrestrial environment, would instantaneously explode in a ball of flame! Whether this turns out to be true of not, any

Fig. 5.1 Titan from the *Cassini* spacecraft (Credit: NASA)

astronaut of the future visiting this strange world will quickly realize that s/he is not in Kansas anymore!

These strange features and interesting speculations will be looked at a little later in this chapter. First however, let us look at where Titan fits in to the scheme of the Solar System.

Titan is, as mentioned in the previous chapter, the largest of Saturn's moons. With a diameter of 3,220 miles (5,152 km), it is also the second largest satellite in the Solar System, exceeded only by Jupiter's Ganymede. It is 50 % larger and 80 % more massive than our own Moon and is actually larger than Mercury, albeit being only half as massive as this dense little planet. The moon is about 1.9 times as dense as water (Fig. 5.1).

Titan orbits Saturn between Rhea and Hyperion at a distance of 764,000 miles (1.2 million kilometers) from the planet. It completes one revolution of its orbit every 15.92 days and has one hemisphere perpetually turned toward its primary. The moon was discovered by Christiaan Huygens on the night of March 25, 1655 but it was almost 200 years later, in 1847, before the name "Titan" was first used (by John Herschel, son of the discoverer of Uranus).

Titan: The Weirdest World in the Solar System?!

For a long time, Titan was thought to be the largest of the Solar System's moons. In fact, it was believed to hold this distinction until the advent of space exploration, when it was toppled from this position by Ganymede. The overestimation of its diameter was due in no small part to a very interesting feature of this object; its dense atmosphere. The presence of this has long been known, but it was not until the moon was examined close up that the extent of this mantle of gas was fully appreciated.

> **Project 1: Visibility of Titan**
>
> The brightness of Titan is variously estimated as between magnitude 8.2 and 8.4. Away from the glare of Saturn, it would be visible without too much trouble in 7×50 binoculars, but what is the smallest instrument that will *actually* show it? As a rule of thumb, it might be thought that if an instrument is sufficient to show the rings of Saturn, it will also show Titan, but is this strictly true? You may already have tried seeing how small an instrument can be and still show Saturn's rings, but did you also notice nearby Titan? Is it easier or harder to find than the rings? What is your experience?

The first inkling that the moon has a substantial atmosphere came as long ago as 1907 when Josep Comas Sola noted darkening near the edges of Titan's disk and reported two roundish white patches—which he suspected to be clouds—near its center. From the rather featureless appearance of most modern images of the moon, it is probable that the white patches were more subjective than real (although a bout of unusually stormy Titanian weather probably cannot be ruled out) but the conclusions that Comas Sola drew were correct. Considering that the disk of Titan is only 0.8 seconds of arc as seen from Earth, these observations are truly remarkable. Subsequent claims of surface markings similar to those seen on Mars were reported by some other astronomers and presented as evidence that the atmosphere of Titan was transparent. We now know that this is not true and, once again, subjective impressions of the very tiny disk were most likely responsible for these alleged markings. Be that as it may, proof of an atmosphere by G. Kuiper in 1944 finally vindicated Comas Sola's conclusion and the moon was officially recognized as being the only one in the Solar System with a substantial gaseous mantle. Methane was detected in the moon's spectrum but the quantity of this gas present was significantly less than the total mass of Earth's atmosphere. On the assumption that this was the main atmospheric constituent, it was concluded that Titan's air was less dense than Earth's.

Speculation as to what conditions on Titan's surface might be like continued, but with little data these were really just guesses. Some of these guesses did nevertheless turn out to be remarkably prescient. For example, Isaac Asimov predicted the existence of lakes of methane on the moon and even raised the revolutionary possibility of the presence of methane-based life. In his typical style, he predicted that the discovery of methane lakes and, maybe, methane life might both come as "titanic surprises" some day. We have already had the first "titanic surprise" as we shall see below.

Following the positive discovery of an atmosphere, further knowledge based upon observation rather than speculation moved little until the *Pioneer* 11 flyby of the Saturn system in 1979. Some low quality images of the moon were transmitted back to Earth by this probe and one relative close-up was secured on September 2 of that year, but apart from showing that the moon had a hazy atmosphere and was very cold (in contrast to some speculations that suggested a Titan warmed by the methane greenhouse effect to almost pleasant temperatures), little was learned. Initial estimates of the temperature suggested that bodies of liquid nitrogen might exist there and artist's impressions of steaming nitrogen lakes under a sky laden with liquid nitrogen clouds purported to represent a typical Titanian landscape. When data revision showed that it was not quite *that* cold, methane lakes and methane clouds replaced the nitrogen ones, but the landscapes remained unaltered.

Titan was visited again in 1980 and 1981 by *Voyagers* 1 and 2. The first of these probes made a close pass of the moon, but observation of surface details was thwarted by the ubiquitous haze. Only much later did intensive digital processing of the images reveal some hints of surface features in the form of a couple of light and dark areas, which by then had actually been found by infrared observations made with the *Hubble Space Telescope*.

The big breakthrough in our knowledge of Titan came on July 1, 2004, when the *Cassini-Huygens* space probe—a joint project between ESA and NASA—arrived in the Saturn system. Its mission included the mapping of Titan's surface by radar and what is probably the most ambitious project yet undertaken in space; the landing of a probe (*Huygens*) on the moon itself.

On October 26, 2004, *Cassini* passed Titan at just 750 miles (1,200 km), securing the highest resolution images of the moon's surface made until that time. Then, on January 14 of the following year, *Huygens* landed on the surface, for a short while transmitting back scenes of a flat area strewn with what appeared to be small bounders but which were in reality lumps of very hard ice. The ground appeared wet with methane and there was clear evidence of flow around the boulders. Moreover, during its descent, the probe gave us unprecedented views of pale highlands crossed by what looked like dark rivers flowing down to a dark plain. Hills, valleys, rivers, deltas and plains all gave a strangely Earthlike feel to the place. Earthlike, yet at the same time weirdly alien as the flows of water responsible for similar features on Earth are simply not possible on Titan (Fig. 5.2).

The main probe—*Cassini*—made another close flyby at just 594 miles (950 km) on July 22, 2006 and another as close as 550 miles (880 km) on June 21, 2010. Thanks to this reconnaissance, the first of Asimov's speculative "titanic surprises"—bodies of liquid—has been confirmed. Titan has several lakes of liquid hydrocarbons; one of them, now known as *Kraken Mare*, is at least as large as Earth's *Caspian Sea* and rivaling this feature as the largest known lake in the Solar System (Fig. 5.3).

As these words are being written, *Cassini* continues its survey of the Saturn system and Titan remains part of this scrutiny. No doubt, some of what we currently think that we know about this moon will eventually be modified or even discarded in the light of new observational evidence, so we must proceed tentatively here.

From what scientists have deduced thus far, Titan is an ice/rock body similar to Jupiter's Ganymede and Callisto. It would seem to be about 50 % water ice and 50 % rock, rather similar to Saturn's Dione and Enceladus, although due to its greater mass, its density exceeds that of either of these bodies. The moon is thought to be differentiated into several layers, each composed largely of

246 Weird Worlds

Fig. 5.2 Lumps of ice are strewn across Titan's landscape in this view from the *Huygens* lander (Credit: NASA)

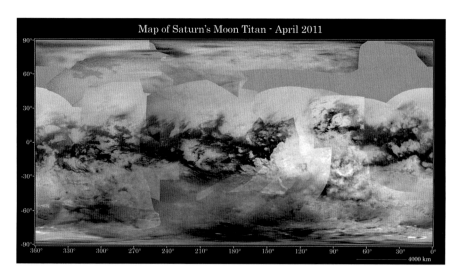

Fig. 5.3 Map of Titan's surface from *Cassini* images (Credit: NASA)

water ice in different crystalline forms, all encasing a rocky core perhaps 1,200 miles (2,000 km) across. There is also evidence for an internal "ocean" of liquid water, or perhaps of an ammonia/water mixture. Several findings by Cassini point in this direction. Extremely-low-frequency electromagnetic waves were detected by the probe propagating through the moon's atmosphere. Because Titan's surface is believed to be a poor reflector of such waves, a more efficient reflector is required to account for this observation and a liquid-ice boundary beneath the moon's surface, such as a subsurface ocean would provide, fits the requirement very nicely. Moreover, *Cassini's* mapping of surface features revealed that these systematically shifted by as much as 19 miles (30 km) between October 2005 and May 2007, suggestive of a crust decoupled from the interior of the moon. This, in turn, suggests the presence of a sub-surface layer of liquid on which the surface "floats". Shades of the "floating" surface of Europa, we might opine!

The prospect of an underground ocean was given a further boost in June 2012 when the results of *Cassini's* monitoring of the extent of Saturn's tidal effect on Titan between February 27, 2006 and February 18, 2011 were published in the journal *Science*. As Titan orbits Saturn, the giant planet's pull causes a degree of deformation; stretching it slightly more when nearer the planet than at greater distances in the moon's slightly elliptical orbit. The extent of this deformation depends upon Titan's inner structure. If this is completely rigid, Saturn's tides would have only a very slight effect, causing a solid tidal bulge of just 3 ft or 1 m. *Cassini* observations show, however, that the bulge is ten times this value; too large to be happening in a rigid and uniformly solid body. These results are nevertheless entirely compatible with the existence of an "ocean" or liquid layer sandwiched between a deformable outer crust and a rigid mantle. Because of the prevalence of water ice, scientists involved in this study consider liquid water to be the most likely candidate for this liquid, although it is likely that ammonia is present there as well. Mixed with water, the latter would act as a form of antifreeze helping to keep the subsurface ocean in a liquid state even at temperatures below the freezing point of pure water.

The question as to whether Titan is tectonically active and whether cryovolcanism occurs there remains unanswered.

In 2004, Argon 40 was detected in the atmosphere and its presence considered by some scientists as supporting evidence for cryovolcanic activity. Nevertheless, direct evidence of this has not been very forthcoming. A hill topped by what looks like a crater (*Tortola Facula*) initially seemed promising, as did a feature found in 2004 and subsequently given the name of *Ganesa Macula*. This latter appeared to be a very volcanic-looking "pancake dome" in early imaging, however further examination of the region in 2008 revealed the "dome" to be nothing more than a chance combination of light and dark patches showing no hint of any volcanic association. Even *Tortola Facula's* volcanic credentials have been called into doubt. Maybe it is simply an odd looking hill with a depression at its summit!

In December 2008, two long-lived "bright spots" were observed in the moon's atmosphere. According to some scientists, these were too persistent to be readily explained as weather events and the suggestion was made that they might have been due to extended cryovolcanic episodes. This is a plausible suggestion, although at our stage in the understanding of Titanian meteorology, we should perhaps be cautious in our interpretation of what is and what is not likely to be a weather event.

Then in March 2009, odd structures rising about 650 ft (200 m) above the surface were found in the region known as *Hotei Orcus*. Oddly, these features appeared to fluctuate in brightness over a period of several months and, although these fluctuations have yet to be satisfactorily explained, the formations themselves have been described as looking very like flows of icy lava. Could they be evidence of cryovolcanism? maybe… but more evidence is needed to decide one way or the other.

Back in 2006, a mountain range 94 miles (150 km) long, 19 miles (30 km) wide and nearly 1 mile (1.5 km) high was found in the moon's southern hemisphere. Mountain ranges of this nature give evidence of something akin to plate tectonics and the suggestion has been put forward that—possibly as the result of a large impact event—tectonic plates were forced to move at sometime in the past, allowing sub-surface material to upwell and form the mountain range.

An alternative hypothesis for this and similar mountain ranges sees them as "wrinkles" on a shrinking world, graphically described as being like the wrinkles of a raisin. (It is amusing that we now have a "raisin" moon joining the "walnut" of Iapetus and the "cracked egg" of Europa. The outer Solar System is beginning to sound like a grocery list!). Titan, according to this hypothesis, is slowly cooling down internally and, as this process proceeds, shrinkage of the outer layers of the moon occurs and parts of the subsurface ocean begin to freeze. According to this hypothesis, it is this shrinking and buckling, not plate tectonics in the terrestrial sense, that pushes up mountain ranges and ridges. Nevertheless, in December 2010, one of these mountain chains, consisting of several peaks between about 3,200 and 5,000 ft (1,000–1,500 m) high, was found to include one very volcanic-looking formation (*Sotra Facula*) in addition to several other mountains topped by what appear to be large craters. Is this evidence of tectonic activity accompanied by cryo-volcanism? As if to provide a further hint, the ground around the bases of these peaks appears to be covered by something that *looks* suspiciously like frozen lava flows.

Nevertheless, "looks like" does not necessarily mean "is". The smoking gun (or steaming mountain!) has not been found and, despite these superficially suggestive-looking clues, many researchers remain unconvinced. For instance, J. Moore of Ames Research Center argues that Titan is not a geologically active world and that all its features can be explained in terms of impact and erosion. Recalling Jupiter's inactive Galilean moon, he describes Titan as "Callisto with weather"! Time and further research will eventually tell who is correct.

An Atmospheric Moon!

One big (though maybe not "titanic"!) surprise delivered by in situ exploration of Titan was the thickness and extent of its atmosphere. The reverse of Mars, estimates made from Earth grossly underestimated the density of the gaseous mantle surrounding this moon. As already mentioned, the presence of methane had been spectroscopically detected back in the 1940s, and in the absence of evidence to the contrary, it was assumed that this gas constituted the bulk of Titan's atmosphere. Space exploration proved, however, that the lion's share of the moon's air is actually nitrogen;

a situation very similar to Earth but unlike any other body in the Solar System. In fact, Titan's atmosphere is even richer in nitrogen than that of our planet, as this gas makes up a whopping 98.4 % of the moon's air. Methane, far from constituting the bulk of the atmosphere, contributes a mere 1.4 %, with the rest consisting of hydrogen (0.1–0.2 %) and trace amounts of ethane, methylacetylene, acetylene, propane, cyanoacetylene, hydrogen cyanide, carbon dioxide, carbon monoxide, cyanogens, argon and helium. Far from being thinner than Earth's, Titan's atmosphere is thicker, having a surface pressure 1.45 times greater and a total mass some 1.19 times larger. This works out at about 7.3 times more massive per equal surface area. Moreover, because Titan's gravity is weaker than Earth's, its atmosphere extends to greater altitudes above the moon's surface. Ultraviolet light from the distant Sun and energetic particles from Saturn's magnetosphere interact with atmospheric methane to produce a cornucopia of hydrocarbons, shrouding the moon in an eternal photochemical smog. This persistent haze succeeds in screening the lower atmosphere from much of the already feeble amount of warmth received from the Sun, creating a so-called anti-greenhouse effect. On the other hand, methane is a strong greenhouse gas and the lower atmosphere therefore compensates to some degree by instigating a (positive) greenhouse effect. However, the net result is still a freezing 90 K or thereabouts at the surface. Yet, given something more substantial than thermal underwear and a heated oxygen supply to keep the precious gas from freezing, Titan is the only place in the Solar System besides Earth where an astronaut could walk about without a space suit! He or she might find walking hard going though; the atmosphere is so thick that standing on the moon's surface would feel like standing at the bottom of a 20-ft deep lake here on Earth!

Another unexpected surprise was *Cassini's* 2004 discovery that the Titanian atmosphere rotated faster than the moon itself; a situation very reminiscent of the super-rotating atmosphere of Venus. Indeed, in certain respects, Titan seems like a low-temperature counterpart of Venus. Because of its slow rotation, there is little temperature difference between equator and pole. This is similar to Venus, but very different to Earth. There is even

Titan: The Weirdest World in the Solar System?!   251

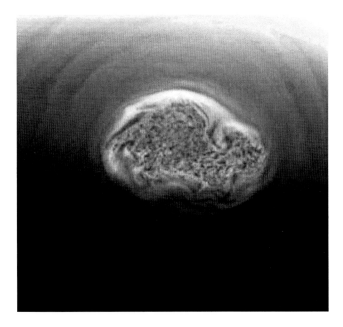

Fig. 5.4 View from *Cassini* of the south polar vortex on Titan (Credit: NASA)

a south polar vortex on Titan; an apparently permanent cyclone reminiscent of the Venusian polar vortices (Fig. 5.4).

In many other respects however, Titan is like a low-temperature counterpart of Earth. Like our home planet, this moon has both static and flowing liquid on its surface. Valleys and channels are carved into highlands by flowing liquid and basins fill to become lakes. The highlands may be made of water ice instead of rock and the rivers and lakes filled with liquid methane rather than water, but the processes remain very similar. Methane is the water of Titan. It fills rivers and lakes, droplets form clouds and, where these are thick enough for droplet coalescence to occur, it falls as rain. In the dense atmosphere and low gravity, the oily drops of methane rain fall slowly, but some of the methane storms can be heavy. *Cassini* saw dry lowland areas turned into lakes following the dispersal of towering cumulonimbus clouds. Methane floods followed the passage of a large storm system and the temporary lakes remaining after the storm may be compared with features known as *playas* here on Earth—low areas which temporarily becomes shallow lakes after rain or flood but dry out into mud

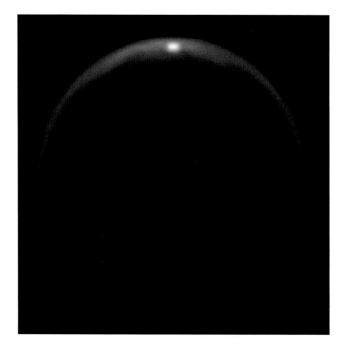

Fig. 5.5 Infra-red image of sunlight reflecting off the surface of Krakan Mare (Credit: NASA)

flats with the return of dry weather. The storm events themselves look very similar to thunderstorm outbreaks on Earth and these storm cells are sometimes referred to as "thunderstorms" in the literature. Nevertheless, lightning has never been observed on Titan. Initially, bursts of radio waves picked up by *Cassini* were interpreted as lightning bolts, but in view of complete lack of any corroborating evidence, these are now thought to have some other cause, possibly interaction with Saturn's magnetosphere. Unlike the *Venera* landers on Venus, which were greeted by a loud clap of thunder, *Huygens* heard nothing. If lightning does occur on Titan, it must be very rare (Fig. 5.5).

At any one time, about 1 % of the moon is under cloud, although cloudy outbreaks covering up to around 8 % have been recorded. These clouds are not to be confused with the ubiquitous smog in which they are embedded. Even if Titan is a far hazier place than Earth, its skies are not frequented by as many clouds.

The Methane Mystery

Methane, as mentioned, is to Titan more or less what water is to Earth. Just as Earth has its hydrological cycle, so Titan has an analogous methane cycle of rain, evaporation and collection in rivers and lakes. And just like Earth's water, its origin is a mystery. Water is thought to have been brought to Earth by "wet" asteroids and short-period comets, but once here it is relatively stable beneath our planet's protective magnetic field. Titan's methane though, is not so stable. Even at the distance of Titan, solar radiation is strong enough to break its molecules apart. The gas is simply not capable of persisting in the moon's atmosphere for more than geologically brief periods of time. Either some source of reproducible methane must exist on Titan or the present abundance of this substance is atypical of its atmospheric composition over most of the moon's existence.

This latter position has been suggested by Kathleen Mandt of the South-West Research Institute in a paper published in the 20 April 2012, issue of *The Astrophysical Journal*. Mandt and her colleagues argue that the present release of methane into the moon's atmosphere began more recently than one billion years ago. Using data from *Cassini's* Ion and Neutral Mass Spectrometer and *Huygens'* Gas Chromatograph Mass Spectrometer, this team of scientists compared the abundance of methane molecules containing carbon-12 with that of molecules containing carbon-13 in an attempt to trace the history of this gas in Titan's atmosphere. The latter molecules are rarer, but they are also heavier and take longer to interact with ultraviolet radiation from the Sun and convert into more complex species. In other words, methane containing carbon-12 gets used up faster, through conversion into more complex hydrocarbons, than carbon-13 methane. By measuring the relative abundance of each, these scientists were able to form some idea of Titan's methane history; and what they found conflicted with the idea that methane had been leaking into the moon's atmosphere throughout its 4.5 billion year lifetime. There was just too little carbon-13 methane! The relative abundances suggested that the methane arrived no more than one billion years ago. It is interesting to note that this result fits a model of Titan's evolution proposed by G. Tobie of the University of Nantes. Tobie suggests that the moon's interior released methane in three bouts during its evolution. The first was around 4.5 billion years ago just after the moon formed, the second about 2.0–2.7 billion years ago and the third between 350 million and 1.4 billion years ago. The methane atmosphere we see today is the legacy of this last period of methane exhalation.

There is, however, another far more speculative and radical possibility concerning the source of Titan's atmospheric methane. On Earth, methane is a byproduct of living organisms and we might recall that it was the apparent discovery of small amounts of this gas in the Martian atmosphere that reignited the debate about contemporary life on that planet. In the same vein, some daring souls have suggested that the methane of Titan might also have a biological source. This is certainly a controversial thought, but as we will see a little later, the possibility of life on Titan is not one to be dismissed too quickly. Moreover, although the hypothesis of Tobie is probably a preferable explanation for the relatively recent release of methane into Titan's atmosphere (in that it involves geological processes and, by the very nature of things, these are simpler than biological ones) it is also true that an evolving biosphere on the moon could likewise explain the increase in atmospheric methane during the past several hundred million years. The atmosphere and biosphere may be co-evolving is a similar way to that of Earth. Although the air of our home planet is rich in oxygen today and although this is a legacy of life, for millions upon millions of years Earth's atmosphere contained no free oxygen, even though the oceans teamed with living organisms. Only with the appearance of organisms whose metabolism produced oxygen as a waste product, did the air of our world begin accumulating this highly reactive gas. Maybe—just maybe—Titanian life has taken an analogous course; existing in this instance for billions of years without seriously disturbing the chemical balance of the moon's atmosphere (maybe confined to the underground ocean?) until around a billion years ago when a methane-producing mutation occurred and altered the moon's history forever. This, it must be said, is not being put forward in very serious tones, but it does present itself as a possibility and should be recalled if all else fails to adequately account for the methane mystery. The broader issue of Titanian life will be taken up a little later. But first, let's take a look at a common feature of the landscape of this strange world.

The Dunes of Titan

Although much of the "fame" of Titan is due to the bodies of liquid found on its surface, the area covered by these hydrocarbon seas amounts to less than 1 %—possibly as little as 0.002–0.02 %—of the moon's surface area. By contrast, somewhere between 12 % and 20 % of its surface is covered by fields of dunes, giving Titan the appearance of a low-temperature version of the fictional planet Arrakis in Frank Herbert's novel *Dune* (presumably, minus the gigantic worms!). Over 16,000 dunes have thus far been noted by scientists monitoring images coming back from *Cassini*. Most of the ones found to date lie within 30° of the moon's equator, although some small fields have been spotted at latitudes as high as 55°. The long equatorial dune ridges make these regions of Titan appear, as one author expressed it, as if they have been "raked" like a Japanese Zen garden. The dunes at Titan's lower latitudes are the longest, narrowest and most closely spaced, indicating a great abundance of sand in those regions. By contrast, those at higher latitudes are shorter and more sinuous as well as being not as closely spaced, indicative of less available sand further from the equator. When first imaged, these latter dune ridges reminded scientists of cat scratches and were initially referred to by this term.

    The occurrence of dunes at all, let alone in the numbers observed, came as something of a surprise. Granted that the moon's atmosphere is pretty thick, it is so far from the Sun that winds strong enough to blow sand into high ridges were not anticipated. Moreover, what exactly is this "sand" that the winds can blow? On a world where ice is bedrock, we certainly do not expect to see the sort of sand that comprises the dunes of Earth's deserts and sea shores. The sands of Titan appear to be made of something quite dark. Furthermore, this material seems to be pretty much ubiquitous, coating the icy surface of the moon in a global layer.

    Combining observations and models, scientists are now fairly confident that they have cracked the mystery of the sand. It consists of either solid particles of complex hydrocarbons or nitriles (i.e. carbon molecules which have an attached nitrogen atom), or (likely) a mixture of both. Unlike the sands of Earth, which are essentially ground up rock, those of Titan are thought to precipitate from the sky. Interaction between the atmosphere, solar ultraviolet radiation and energetic particles from Saturn's magnetosphere synthesizes complex organic materials in the moon's upper atmosphere, creating the everlasting smog that perpetually shrouds its face. Slowly however, these particles drift toward the surface and settle as a light organic snow, over time building up into vast deposits. Probably, sedimentary layers of these deposits exist and are subject to erosion by methane rainfall and flowing channels of liquid runoff. At this point, the genesis of terrestrial and Titanian sand shares a common process—erosion. Liquid methane flows, wind transportation or both conspire to carry these particles to sand sinks, from whence global winds eventually blow them into dunes. By the way, it is estimated that there is more organic material in the dunes of Titan than in the whole of Earth's coal deposits!

Now dunes—whether on Earth, Mars, Titan or Arrakis—are interesting things in their own right. Unlike hills and mountains, they move at what is really a rapid rate compared with the slow changes of most other landforms. The process by which they form and move is known as *saltation*; involving what is essentially a "jumping" of individual grains. This has been well studied over the years both in the field and in wind tunnels and it is known that a combination of sufficiently strong winds and dryness of sand is necessary for the grains to jump forward in the way required. As each sand grain jumps forward and hits the ground, it kicks up others from the surface, thereby recruiting them into the forward march. The process occurs in a thin layer near the ground and can become self-sustaining. If there is some irregularity in the ground, the moving sand will tend to pile up around it and start a dune growing. Once it starts, a dune may exist and keep on growing for a long time; sometimes even exceeding a 1,000 years.

The largest types of dunes—known as *linear* or *longitudinal* dunes—form when the wind blows predominantly from two different directions. The net sand transport effectively takes the mean between the two directions and the dunes can stretch for great distances aligned to an average between the directions of the two most commonly prevailing winds. The dunes of Titan are of the linear variety. By studying their orientation, scientists can derive the direction of the moon's prevailing winds. However, the initial results turned out to be yet another surprise! The orientation of the dunes seemed to suggest that the prevailing winds blew from west to east. The problem was, models of Titan's atmosphere based upon the rotation of the moon, indicated that they should be blowing in the opposite direction—from east to west. Ouch! So counter to expectations did the results appear, that atmospheric scientists initially wondered if the images had flipped the wrong way around! The *Huygens* lander indicated that the winds close to the equator at the time of its descent to the surface were blowing from the north-northeast. This did not help to solve the problem!

A new model of the moon's atmosphere developed by T. Tokano of the University of Cologne and R. Lorenz from the John Hopkins University Applied Physics Lab does, however, offer a way out. These scientists took into consideration the likely wind history of the moon over its entire 30-year solar orbit and concluded that the predicted east-to-west winds are indeed the prevailing ones. Nevertheless, for brief periods around the equinoxes, stronger winds blow in the opposite direction and, although these west-to-east winds are rare, their velocity alone dictates that they are more efficient at moving sand particles and sculpting the dunes. Not that these winds are necessarily strong by terrestrial standards. All that is required to move the sand are breezes of between 2 and 5 miles (about 3–8 km) per hour, although this does not mean that velocities may not at times exceed these values. Moreover, because of the thickness of the Titanian atmosphere, a gentle breeze packs more of a punch than one of equal velocity on Earth.

So the mystery of the Titanian dunes appears to be more or less cracked. One puzzle solved, but how many more to go?

Is There Life on Titan?

The time has come to take a closer look at what might turn out to be the weirdest thing of all about this large Saturnian satellite; the possibility (and it is *only* a possibility we must remember) that it harbors life. Not just a few simple cells in atypical environments (as might be true of several other Solar System bodies as already discussed) but that it might be *teeming* with life and that this life may be of a kind unlike anything found on Earth!

Starting at the conservative end of these speculations, evidence for a subsurface ocean of water or water mixed with ammonia has led to the same sort of speculation concerning simple aquatic life that has been raised about several Solar System denizens such as Europa, Triton and even the dwarf planet Ceres. Evidence for a Titanian ocean deep underground is quite strong, as already noted, but whether life could have emerged there is another matter. Some scientists raise concerns as to the availability of the required minerals in an ocean that is sandwiched between layers of ice and not in contact with appreciable quantities of rock, but in our present state of ignorance we can make no firm judgment as to what might or might not be dissolved down there. As with any extraterrestrial subsurface ocean however, proving or disproving the presence of life therein will be no easy task and getting a probe into the Titanian ocean will make the probing of Europa's equivalent feature seem simple by comparison!

(Maybe) "Its Life Jim, But Not As We Know It!"

Once the nitrogen/methane composition of Titan's atmosphere was firmly established, astrobiologists' interest in the moon grew even keener. They saw it, not so much as a home of life today, but as an environment having enough similarities with early Earth to maybe become a home for life in the distant future. Titan seemed like a good "pre-biotic experiment" whose study might throw light on the pre-biotic Earth. This became even more pertinent following experiments, by Professor Roger Yelle and University of Arizona and graduate student Sarah Horst, in which gaseous mixtures duplicating the upper atmosphere of Titan were irradiated in an attempt to reproduce the sort of reactions expected to follow bombardment by energetic particles from Saturn's magnetosphere and ultraviolet radiation from the Sun. Amongst the complex organic molecules formed were the amino acids glycine and alanine as well as nucleotide bases including cyctosine, adenine, thymine, guainie and uracil; the five bases used by terrestrial life!

These substances are presumably numbered amongst the mixture of organic compounds that continually waft downward through Titan's atmosphere, eventually coating the moon's surface and falling into the liquid hydrocarbon lakes and seas.

This is interesting in its own right, but several astrobiologists have gone even further than proposing Titan as a pre-life environment, or even as an environment that might just conceivably harbor simple life-forms, more or less akin to the single-celled organisms found in stagnant ponds on Earth. They suggest that the moon may be the site of a form of life totally alien to anything yet studied; viz. methane-based organisms. Lifeforms of this type are not simply the cousins of terrestrial methanogenic bacteria like the hypothetical organisms proposed by some scientists to explain the methane found on Mars. Methanogens on Earth—and on Mars as well if they really do exist there—still need water as a solvent and are in that respect no different from trees, birds and human beings. But what is proposed as a possible form of life on the surface of Titan are organisms for which liquid methane replaces water as a solvent!

The postulation of methane as a biologically-suitable solvent is a daring and controversial one as nobody knows whether it really is capable of playing such a role. On the one hand, it is not as strong a solvent as water but, on the other, it is also not as reactive. Water can break down complex organic molecules through the process of hydrolysis in a way that methane does not, so in that respect at least, methane appears to have some advantage over water as a "solvent of life". On the downside however, liquid methane environments are necessarily very cold places, so any living organism found there will need to be very efficient at converting its food into energy. Even so, metabolism—assuming that it could function at all under these circumstances—would probably be very slow compared with that of most terrestrial organisms. Whether this sort of metabolism is possible at all is the big unknown question, but if it is, the "energy efficiency" of such an organism would be so high that if it was brought into contact with the environment in which terrestrial life happily functions, it would explode in a ball of flame! Where water based life and methane based life are concerned, the twin can surely never meet!

Now, *if* such life truly does exist, the implications for exobiology are enormous. For a start, the widespread occurrence of methane on Titan would probably allow methane-based life to cover the moon, just as the presence of water permits life in some form to be found virtually everywhere on Earth. But, in the wider scene, the universal boundaries of life would be tremendously expanded. There are likely to be many more Titan-like worlds in the Universe than Earth-like ones and therefore if there are such things as methane-based organisms, they may well be the "normal" expressions of life in the Universe. Water-based life of the type known on Earth might be the anomaly. As for Titan itself, should it indeed be true that water-based organisms inhabit a subsurface ocean and methane-based ones live out their existence on the moon's surface, this small world becomes home to two totally diverse and unrelated forms of life; each alien to the other, each unable to exist in the other's environment and each presumably hailing from a totally separate genesis. Biologically speaking, Titan could be two completely alien worlds in one! Now that would be truly weird!

But of course, all of this remains little more than science fiction unless methane-based life is shown to be possible, and it is here where some speculative but scientifically disciplined reasoning by a group of astrobiologists earlier this century throws down the really interesting challenge. Their work makes methane-based life appear possible "on paper" and opens up the prospect of further observations of Titan either strengthening or weakening the prospect of its existence in the real Universe as well.

In 2005, Helen Smith and Chris McKay published a paper proposing a hypothetical form of life that might theoretically exist on Titan or similar worlds. They proposed methane-based organisms that breathed hydrogen and consumed organic compounds such as acetylene and ethane. The energy needed by these hypothetical organisms would be supplied by combining acetylene with hydrogen to produce methane, the latter being expelled back into the air as a waste product of the metabolic process. A variety of organic compounds might supplement the diet of these organisms; and if we can be sure of one thing about Titan it is that this moon abounds in organic material!

Because such life, if it exists at all, would probably be widespread across the Titanian surface, Smith and McKay suggested that it should be sufficiently abundant to leave its mark on the atmosphere of the moon, taking it out of chemical equilibrium in certain respects, just as terrestrial life throws our own atmosphere out of equilibrium (the abundance of that very reactive gas called oxygen being the chief case in point). The telltale influences of the sort of life they proposed would be different to those of terrestrial life of course, but they should nevertheless be detectable. Working from their model, they concluded that the existence of this form of life might be betrayed by a significant lack of ethane and acetylene on the surface of Titan compared with that predicted from purely abiotic models of the moon's atmosphere. In particular, photochemical processes should mean that ethane is especially abundant. Models predicted that the Titanian surface should be covered by a layer of ethane many meters deep. Moreover, because these hypothetical organisms are supposed to be hydrogen breathers, their presence should mean that hydrogen should be depleted at the surface, mysteriously disappearing as it is consumed and combined with organic molecules to form methane.

How have these predictions shaped up in the light of continuing analysis of the moon by the *Cassini* and *Huygens* probes?

Remarkably well as it turns out. For a start, acetylene *is* seriously depleted. None was observed amongst the gases released from the surface during the *Huygens* landing and continuing observations have confirmed that it is seriously below predicted levels. Similarly, Roger Clark and colleagues found that ethane is also strongly depleted compared with predicted quantities. Neither ethane nor acetylene is widespread on the surface, strongly contradicting predictions, but plenty of benzene has been found there, accompanied by an unidentified substance which appears to be some complex organic molecule that has, to date, eluded identification. These findings are based upon direct observation and are considered beyond dispute.

A potentially even more startling result however is the apparent discovery by Darrell Strobel of a flux of hydrogen into the moon's surface and its subsequent disappearance. Models of the moon's atmosphere predict the presence or hydrogen due to the breaking down of molecules of methane and acetylene through the action of ultraviolet radiation. These models also indicate that the hydrogen will be evenly distributed throughout the depth of the atmosphere, but Strobel found that the predicted distribution is not consistent with data from *Cassini's* infrared spectrometer and neutral mass spectrometer. What he found was an apparent disparity in the lower atmosphere suggesting a downward flow to the surface, and subsequent disappearance into the surface, of some 10,000 trillion trillion (the number 10 followed by 24 zeros!) hydrogen molecules every second. The number that seems to be disappearing into the surface is about the same as that escaping from the upper atmosphere. Hydrogen wafting upward and escaping from a small world such as Titan is not a problem. Hydrogen travelling in the opposite direction and then vanishing at ground level is! As Strobel picturesquely expressed it, "It's as if you have a hose and you're squirting hydrogen onto the ground, but it's disappearing." He does not think that there is some sort of "hydrogen storage" under the surface of Titan and notes that the very low temperatures of the environment makes it unlikely that hydrogen is reacting with some other substance unless a type of catalyst capable of operating at these temperatures is available on the moon's surface. Strobel notes how his results correspond with the predictions of McKay and Smith, but also stresses that a mineral catalyst at the moon's surface would equally explain them. Whilst not as exciting as finding life, the discovery of something capable of acting as a catalyst at the temperatures found on Titan would be a significant discovery by itself.

This result, unlike acetylene and ethane discoveries was, however, not derived from direct observation but from a computer simulation designed to fit measurements of hydrogen in the lower atmosphere. As such, it is not considered to be as secure as the acetylene and ethane results, but if it is confirmed it would strongly suggest that the observed depletions are not simply the result of lack of production of these substances. It would be strong evidence that something is actively destroying them at the surface, although it would not automatically prove that this "something" is biological. It would, however, mean that we are faced with two equally weird explanations—either an unknown catalyst or an unknown form of life. Either solution hints at a "titanic surprise" of one form or other.

Nevertheless, when taken all together, these results appear to confirm the predictions of Smith and McKay. It is not easy to see why acetylene and ethane are not being produced according to atmospheric modeling, nor is there any obvious reason to suspect that, once produced, they are being destroyed at the surface. Still, Occam's Razor and caution dictate that all non-biological possibilities must be looked at before the very radical hypothesis that this is not just chemistry, but a totally alien form of biology, is seriously entertained. For instance, Mark Allen, principal investigator with the NASA Astrobiology Institute Titan Team suggests that the deficiency of acetylene might result from the action of ultraviolet light or cosmic rays transforming atmospheric acetylene into more complex aerosols that settle onto the surface but no longer show the acetylene signature. Further atmospheric modeling, and maybe laboratory experiments reproducing Titanian conditions, will presumably sort out whether such suggested simpler processes can account for these strange results or whether investigators will eventually be required to bite the bullet and propose alien life as being the most likely explanation. Should that happen, we can be pretty sure that a new Titan probe with sophisticated means of biological detection will rocket to the top of the list of proposed space missions in the hope that thorough on-site investigation will settle this important question one way or the other, once and for all.

One might raise the argument that nothing suggesting life was apparent on the surface images sent back by *Huygens*. Admittedly, this was only a brief view of a very small region of the moon's surface, but as a damp lowland area it was more or less equivalent of the regions of Earth where we might expect to find lush vegetation. Nothing of the sort greeted the robotic eyes of the *Huygens* probe but that does not necessarily rule out life at that site. If there is life on Titan, chances are that it is microscopic. Most life on Earth is microscopic and many astrobiologists believe that if life does exist beyond Earth, most of it will be in the form of microscopic single-celled organisms. In short, the Titanian scene spread out before Huygens could very well be teeming with living organisms without giving any visible evidence, just as a stretch of fresh soil on Earth may reveal no sign of the millions upon millions of soil bacteria who call it home. Even larger organisms might be present and still escape detection. Look across a ploughed field here on Earth and tell me how many earthworms you can see!

None presumably, but we all know that this does not mean that the soil is free of them.

Cassini and *Huygens* are not likely to prove or disprove the existence of Titanian life. Maybe a future probe will, although there are no firm commitments for anything of the sort in the near future (although if other and stronger hints of life's presence are found, that might change!). But whether there is life on this moon or not, the presence of a world within our home planetary system that is simultaneously so like our own and yet so weirdly alien gives us a hint of what might lie further afield. If a sister world can be this weird, what may lurk in the depths of the Galaxy, orbiting other suns that may in themselves be very different from our own? Let's take a look at some of these far flung worlds and see what awaits us there.

6. Weird Worlds Far Away

This may not be the usual way of setting out a chapter, but this one begins with an "aside". If the reader wonders where all this is leading, the purpose will hopefully become clear in a few pages time when we launch into the real subject of this final chapter.

Copernicus and "Non-special" Earth in a Universe of Weird Worlds

A curious ambiguity has haunted the way we have thought about Earth in relation to the rest of the Universe. Ever since the Copernican heliocentric model of the Solar System became accepted, this factual discovery has been made to carry the weight of a philosophical interpretation that has had both positive and negative effects on scientific progress. Who has not heard of the "Copernican Principle"? Nobody who has ever picked up a book popularizing science I would wager! The canonical interpretation runs something like this: Prior to Copernicus, people thought that Earth was at the center of all things, that the Universe was small and "local" and that (simply by being at the center of it all) Earth possessed special importance. A second strand of this pre-Copernican world view (though one stressed less in popular-level books) was the assumption that "terrestrial" and "celestial" physics was radically different. The Earth was composed of one kind of matter and the astronomical bodies of another, with the orbit of the Moon representing the point of demarcation. "Sublunary" implied the Earthly realm and "superlunary" the realm of the celestial bodies. This was the old system, handed down from Aristotle through Ptolemy and blessed by the Papacy as unquestioned orthodoxy.

Then, Copernicus came along and presented an alternative model which dethroned Earth from its central position in the cosmos to just another of the planets orbiting the Sun. Subsequent advances in scientific knowledge have continued and extended this process of dethronement; removing, first, the Sun from its central position in the Universe and eventually consigning even the Milky Way galaxy to just one of what might be an infinite number of similar systems throughout a Universe that effectively—and perhaps literally—extends forever. It is this "dethronement" of Earth and its cosmic environment (Solar System, Galaxy etc.) that has been enshrined in the "Copernican Principle". This principle can be summed up briefly as the recognition that our (Earth's) position in the Universe is not in any way special. We occupy no "privileged" or "special" position in the scheme of things, contrary to what the pre-Copernicans are said to have held.

As a rough overview, this might be a fair enough assessment, but it does enshrine some oversimplifications and misinterpretations. For instance, a closer look at the sublunary/superlunary dichotomy uncovers a very different interpretation of Earth's alleged "central" position in the Universe from the one that is popularly promoted. Indeed, the Aristotelian/Ptolemaic model placed our world in a far less salubrious position than is frequently supposed. Far from being seated on the throne of the Universe, this cosmological model saw Earth as being the place where "gross" matter accumulated. It shared none of the glorious qualities of the stars and planets and was effectively the sump of the Universe. It may be going too far in the opposite direction, but we could say that, far from seeing humankind as princes occupying the cosmic throne, this model saw us as rats in the cosmic sewer! Far from dethroning Earth, Copernicus actually *elevated* it to the level of the celestial by merging the sublunary and superlunary into a single cosmos.

This "elevation" of Earth later became a specific aim of one of Copernicus' most famous disciples—Galileo. Explaining the phenomenon of earthshine on the crescent Moon's darkened disk, Galileo stated that "I will prove that the Earth ... surpasses the Moon in brightness, and that it is not the sump where the Universe's filth and ephemera collect". No clearer statement of the "elevating" potential of the Copernican model could be required!

The widely-held notion that pre-Copernicans thought of the Universe as small and "local" also needs a drastic rethink. Part of this notion comes from an old drawing showing Earth as a flat landscape surrounded by a thin film of stars through which an intrepid traveler pokes his head to view the mechanisms beyond. No doubt, the reader will have seen this in any number of books on the history of science, where it is typically referred to as a mediaeval drawing depicting the beliefs current at that time. In fact, it dates from the 1800s and depicts what somebody *thought* the mediaeval beliefs to have been! In fact, the Ptolemaic tradition was quite aware that the Universe was vast and the stars distant. Ptolemy dealt with the subject in Chap. 6 of the First Book of the *Almagest* where he wrote that "the Earth has sensibly the ratio of a point to its distance from the sphere of the so-called fixed stars". To say that something has "sensibly the ratio of a point" compared to the distance of some other object implies that, for all practical purposes, the distance of that other object is infinite.

The Copernican model cannot therefore be presented as the radical dethronement of our position in the Universe in the way that is often claimed. Copernicus' heliocentric model of the Solar System and its subsequent verification represented one of many advances in scientific knowledge over the centuries, but the perception that it was also a radical break in humankind's view of its place in the scheme of things has also led to its enshrinement as a philosophical principle as well. This enshrinement (itself based upon an exaggerated appreciation of Copernicus' break from the past) has been a mixed blessing for science. The (philosophical) Copernican principle, which basically assumes that there is nothing special about our place in the Universe, is a good working principle for investigating the cosmos, but it has also led scientific thought astray on some important issues. For instance, it was the assumption that Earth was typical of other planets that inspired the assumption that these must also be inhabited. Why not? If Earth is inhabited and if it is in no way special, then life (even human life) should be the rule and not the exception. The argument for intelligent Martians flowed directly from this line of reasoning.

More generally, Earth was taken as the model for all planets in the sense that these were expected to be physically similar in their basic form to our own. Giant planets such as Jupiter were imagined as bigger versions of Earth, having the same general features such as well defined atmospheres, solid surfaces and the like. Paradoxically, Earth was treated as being "special" in virtue of being commonplace; the broad archetype of what any planet should look like!

We now know that Earth is not the model for these worlds. In fact, most of the known planets are not very Earthlike at all, and even those that are basically similar to our own still differ from it in some important details.

In a wider sense, a group of cosmologists in the middle years of last century argued—entirely logically—that if the Copernican principle (more precisely, the Cosmological Principle which is basically a more developed version thereof) is correct insofar as there are no privileged points in space and if time can be represented as a fourth dimension of the space-time continuum a la Einstein, it should follow that there are no privileged *moments* in time either. Of course, the Big Bang represents the most "privileged" of all moments and so, on this argument, violated the Principle most fundamentally. From this was born the Steady State theory of the Universe.

Well, we now know that the other planets of the Solar System are not inhabited by human-like creatures and that worlds such as Jupiter have little in common with Earth. The Steady State theory of the Universe has faded into the history of thought. But prior to the 1990s, the Copernican principle continued to hold sway over our ideas of extrasolar planets and other solar systems. Not that this was a conscious thing. The Copernican principle has become such an unconscious assumption that we just think in Copernican terms as naturally as we breathe the air. If there are other solar systems in the Galaxy, of course they would be like our own! To think differently would imply that ours was somehow "special" … and that would be un-Copernican and therefore heretical!

But once again, the Universe has shown that it does not slavishly follow the Copernican rules that we attempt to impose upon it. Hundreds of solar systems have now been discovered, but not one bears much resemblance to our own and some are so alien as to be truly bizarre.

In this chapter, a quick tour of these weird systems and the weird planets that occupy them will be taken. If some of the worlds in the Solar System appear strange according to our standards, we haven't seen anything yet! And because the investigation of extrasolar worlds is such a new field, we can be sure that we have only scratched the surface of what really does lurk out there. What the future will uncover beyond the Sun can only be guessed.

Early Hints That Came to Nothing

The existence of planets orbiting stars other than the Sun has long been accepted by astronomers, albeit more as an article of faith than as the result of specific observations. It was simply seen as a consequence of the hypothesis that the Sun and planets formed from a disk of gas and dust as postulated long ago by Kant, Laplace and Swedenborg. If stars formed in this way, planetary systems would be almost inevitable consequences of the process and should surround every star in the sky, excepting the odd instance where some event may have prevented the process from going to completion. The alternative theory of the Solar System's origin (viz. that the Sun's planets condensed from a stream of gas pulled from the Sun by the tidal effect of a second star passing at very close range) would make planetary systems rare occurrences, but given the billions of stars in the sky, very close encounters must take place from time to time and the total number of planetary systems would still be large. However, with the weight of evidence increasingly coming down on the side of the "nebular hypothesis" of Laplace and Co. the existence of an abundance of extrasolar planets appeared certain.

Nevertheless, actually finding planets around other stars would not be easy, for reasons too obvious to require comment. Prior to recent years, visual detection was out of the question, however it was recognized that a large planet orbiting a nearby star might be tracked down through its gravitational perturbations of the star, not unlike the way that Neptune was tracked down through its effect on the orbit of Uranus. If a large planet was associated with a double or multiple star system, it might give its presence away by influencing the motion of the component stars as they orbited a mutual center of gravity. The effects would be far less obvious than Neptune's perturbations of Uranus, but a large enough planet may just be detectible. Moreover, even a planet of a single star might be detectable if it caused enough of a wiggle in that star's proper motion to be apparent after several years of accurate monitoring. Needless to say, only relatively nearby stars, those having sufficiently large proper motions, could betray their planetary companions through this method and only the very largest planets would stand any chance of detection.

A suspected planet of the binary star 70 Ophiuchi was announced by W. S. Jacob at Madras Observatory in 1855 as a probable cause of "orbital anomalies" he claimed to have detected in the system. Later—in the 1890s—T. J. See of the United States Naval Observatory made a similar claim for this star and even deduced an orbital period around one of the stars of 36 years for the supposed planet. Nevertheless, calculations by F. Moulton cast severe doubt on the existence of such a planet, showing that the proposed system would be unstable if the planet really did move in the orbit derived by See.

Several claims of planet discoveries were made by P. van de Kemp and his team during the 1950s and 1960s, most famously the alleged discovery of two Jupiter-like worlds orbiting Barnard's star. This nearby red dwarf has a large proper motion and appeared to show definite, regular, deviations in its track across the sky. The existence of the Barnard's star system was widely accepted for years, but unfortunately the telltale wiggles in the star's proper motion failed to be detected by teams working at other observatories. Eventually, the real cause of the deviations was found to lie in a problem with the telescope that van de Kemp was using, not with the motion of Barnard's star itself. The other reported discoveries claimed from observations made with the same telescope naturally fell under the common cloud and are mostly rejected today.

Only one relatively early planetary suspect appears to have withstood the test of time. This was the possible object that a Canadian team of astronomers, B. Campbell, G. Walker and S. Yang, suspected to be in orbit around Gamma Cephei. This was eventually confirmed, albeit not until more precise observational techniques became available in 2003.

More recently, in 1991, A. Lyne and colleagues noted apparent variations in the pulses of the pulsar PSR 1829-10 which they initially suspected as being due to a planet orbiting this collapsed stellar remnant. This interpretation was soon withdrawn, however the strange notion of planets orbiting pulsars which these astronomers briefly raised, albeit incorrect in that instance, did not go away. They truly had raised the curtain—admittedly prematurely—on a whole new class of hitherto unsuspected and truly weird planets! Let us look a little more closely at these planets and the strange stars which they orbit.

Planets of the Dead Stars

When a large star reaches the end of its life, it does not (in the words of Dylan Thomas) go gently into the night. Raging against the dying of the light, it blazes for several months in one of the most violent explosions we can witness; a supernova. There are several varieties of supernovae, but the ones of concern here are of the so-called core collapse kind in which a massive star uses up its internal fuel and implodes. Oversimplifying the process, we know that stars gain their energy from the thermonuclear fusion of elements—hydrogen into helium for most stars during most of their lives. Fusion of hydrogen into helium gives out a lot of energy; far more than is required to get the process underway. This is notoriously demonstrated by the hydrogen bomb which requires a small fission device to trigger an explosion beyond anything that a fission bomb can achieve on its own. This is a bad thing for us in the case of the H-bomb, but a good thing with respect to the Sun. But as we go to heavier and heavier elements, more energy is required for the reactions to proceed and less is produced once they do get started. Small and medium stars simply cannot conjure up enough power for the heavier elements to fuse in their interiors. The more massive ones however, are capable of driving fusion all the way to iron. But it is this very capacity for making iron that ultimately dooms them. The problem is, the energy needed for the production of iron as a fusion product equals the amount of energy that this reaction produces. Now, for a star to maintain equilibrium, the energy of its internal thermonuclear reactions must be sufficient to counter the force of gravity ever acting to implode it. This balance between gravitational collapse and disruption by radiation pressure is what keeps a star stable. As radiation pressure puffs it out, the core temperature tends to fall, the reaction slows and gravity starts winning the struggle. The star begins to collapse, heats up at the core, thermonuclear reactions rev up again, radiation pressure increases and equilibrium is maintained. (This is all grossly oversimplified but, I hope, captures the basic principle). However, as more and more of a massive star's core is converted into iron, more of the energy formerly available to keep gravity in check is required just to keep the fusion reaction going. At some point, gravity wins—spectacularly! The star implodes, crushing the core into an object no larger than the central business district of a not-very-large city, but having a density equal to that of an atomic nucleus. Electrons and protons are crushed together in the very atoms of which it is composed. The rebound from this collapse rips through the star, setting off fusion in the "unprocessed" material within the body of the star and, to put it briefly, blowing the whole thing to kingdom come. What remains after the mighty explosion is the star's core—a neutron star—surrounded by a steadily attenuating cloud of gas. These neutron stars spin incredibly fast, have immense magnetic fields and betray their presence by shooting out beams of radiation like high speed celestial lighthouses. We know them as *pulsars*.

Common sense tells us that something born in fire and destruction of this magnitude could not have a planet in orbit around it. Nevertheless, common sense in this instance has been proven wrong!

About a year following the pulsar planet false alarm mentioned above, radio astronomers A. Wolszczan and D. Frail announced that they had discovered two planets orbiting another similar object, this one designated as PSR B1257+12. The planets revealed their presence by causing very slight variations in the pulse as the pulsar wobbled around the mutual center of gravity of the planetary system; a slight but nevertheless detectable effect.

Not surprisingly in view of the false alarm the previous year, many astronomers were initially skeptical about the reality of these planets. Adding to their skepticism was the fact that a pulsar was about the last place where one might expect to find planets. How could a planetary system orbiting a star that subsequently exploded as a supernova possibly manage to survive the blast? Nevertheless, follow-up observations confirmed the "impossible"; this time, the pulsar planets were real. Not only that, but a third planet was later found in the same system.

The three planets in orbit around this pulsar—rather prosaically designated as PSR B1257+12 b, c and d—orbit their exotic sun in the same plane, not too unlike the orbs of our own System. The big difference (other than the obvious one that their central sun is a pulsar and not a star of the Main Sequence) lies in the size of the system. The orbits of all three could fit within the orbit of Mercury and each has a "year" of just 25, 66 and 98 days respectively. With respect to size, two of them may be thought of as Earthlike in a very general sense (4.3 and 3.9 Earth masses for the outer pair), but only 0.02 Earth masses for the innermost. This latter currently holds the distinction of being the least massive extra-solar planet yet discovered and there has been some thought that it might better be known as the first detected extra-solar asteroid or, at the least, dwarf planet. But it is probably safest to stick with the "extra-solar planet" label for the time being!

In addition to these three planets, the pulsar is also suspected of having something else orbiting at much greater distances. The possibility of more remote objects has been raised on a number of occasions, but the suggestions as to what these objects might be have covered a wide range of possibilities. An early suggestion involved a large planet of roughly 100 Earth masses orbiting the pulsar at approximately the distance of Uranus once every 170 years. At the other end of the scale, a tiny asteroidal body just 0.0004 times the mass of Earth in a 3.5-year orbit around 2.6 times the Earth/Sun distance was proposed. A sublimating body like a comet was another proposal and an asteroid belt out beyond the equivalent distance of Mars, yet another. This strange system, it would seem, has yet to give up all of its mysteries.

The big mystery is how these planets came to be there in the first place. Two hypotheses have been proposed. The first puts forward the thesis that the planets are all that remains of Jupiter-like bodies that formed together with the star, rather in the manner of the Sun's planetary system. Prior to the supernova explosion, these were "normal" gas giants and probably orbited the star at significantly greater distances. The blast, however, succeeded in stripping away most of their mass, leaving only core remnants whose orbits contracted closer to the pulsar where we find them today.

The other alternative sees the planets as being second-generation objects that formed after the supernova explosion occurred. According to this model, any original planets which may have orbited the star during its Main-Sequence lifetime were either destroyed or set adrift from the system when the star went supernova. However, it is hypothesized that after the supernova explosion occurred, material pulled from a companion star by the pulsar formed an accretion disk around the latter and it was from this that the planets condensed. In a strange sort of way, these two hypotheses mirror the two models of the formation of the Solar System; albeit with some further additions!

Either hypothesis is possible. The fact that the orbits of all three planets are confined to essentially the same plane strongly suggests that they formed from an accretion disk, but it does not tell us whether this disk was a first or second generation feature.

In whatever manner they formed, these planets are truly weird, at least by the familiar standards of Earth. Being so close to the pulsar, they are perpetually bathed in intense radiation, although whether they are exposed to the full dose from the powerful beams is not known. Probably, the radiation bombardment (even if they do not receive the full force of the pulsar beam) is slowly evaporating the planets, as the solar wind is doing to Mercury in a milder way. If this erosion is sufficiently severe, the atoms sputtered from the surface might give the planets a Mercury-like atmosphere and, if this is dense enough, excitation of the atomic particles by the pulsar's radiation could result in a perpetual "auroral" glow. It has also been suggested that X-rays from the pulsar might cause the surface of the planets themselves to fluoresce, so the hemisphere turned toward the pulsar may be literally glowing as well as surrounded by the pale glow of the thin aurora-illuminated atmosphere. All of this is, it must be said, speculative though certainly not fanciful.

There is a good possibility that the planets are tidally locked so that one side perpetually faces the pulsar. The side turned away would then be in perpetual darkness and incredible cold. Even the hemisphere facing the pulsar is not exactly in daylight. Although pulsars are not "dark", they do not give out much light either and even from the innermost planet of this system, the stellar ember is so small and faint as to appear as little more than a really bright star, estimated to be about half as bright as the full Moon appears to us.

This pulsar is not the only one known to host planets. A millisecond pulsar (catalogued as PSR B1620-26) located some 5,600 light years away in the Scorpius globular star cluster M4, is also known to be orbited by a planet; however this system is very different from that of PSR B1257+12. For one thing, the pulsar is one component of a double star, the second being a white dwarf whose distance from the pulsar is about the same as that separating Earth and Sun. These two decayed stars orbit a mutual center of gravity, and beyond the pair—at a distance roughly equal to that of Uranus from the Sun—a large planet completes its orbit of the two stars once every 100 years or thereabouts. It is widely held that this was originally a planet of the proto-white dwarf and, if it truly formed together with that star, it is the oldest planet known. The age of the star is around 12.6–12.7 billion years. Not surprisingly, although its official designation is the rather prosaic PSR B1620-26b, it has been unofficially dubbed *Methuselah* and *Genesis Planet*. Finding a planet of that age was quite a shock actually. Most theories of planetary formation require a higher concentration of heavy elements (or "metals", in the eccentric astrophysical use of that word) than one would expect to be available to stars that formed upward of 12 billion years ago, at a time when the chemical evolution of the Universe had not proceeded far enough to enrich the interstellar medium with significant quantities of these elements. In view of this difficulty, it has been suggested that the planet is actually far younger than the stars around which it orbits. Maybe it formed from the more enriched "ashes" of the supernova that produced the pulsar. Or, just possibly, it did not begin life as a planet at all, but is really the core of a small star remaining after the outer layers were blown away by the supernova explosion. It is also just possible that this was once one of the "gypsy" planets that we shall meet later, but became captured by the star system as what we might call an "irregular" planet, by analogy with the captured moons of the giant planets of the Solar System.

Whatever the truth about PSR B1620-26b, the "stripped star" process given above as one of the explanations for its presence is accepted as the most likely explanation for another planet found orbiting a second millisecond pulsar; PSR J1719-1438. This planet whips around the pulsar every 2 h as it pursues an orbit just 375,000 miles (600,000 km) from the pulsar itself. The radius of this orbit is slightly less than that of the Sun! The planet itself is estimated to be about half as big as Jupiter and to be some four times as dense as Earth, giving it the highest average density of any known planet. It is at least as dense as platinum although nobody suggests that it is composed of that element!

Now, millisecond pulsars are thought to begin as ordinary pulsars originally constituting one component of a close double star; a component that went supernova. Thanks to the extreme tidal pull of the pulsar, gas is pulled from the companion star, reducing its mass while increasing that of the pulsar and spinning up its speed of rotation until it whirls on its axis at a rate in excess of 10,000 times every minute. It only stops feeding off its companion when the latter is reduced to little more than an ember—often a white dwarf as in the case of PSR B1620-26. However, in the instance of PSR J1719-1438, the companion star was *so* close to the pulsar that it lost all of its outer layers and all of its lighter elements, leaving only a comparatively tiny core remnant containing just 0.1 % of the star's original mass. Because the lighter elements were stripped away, this remnant consists predominantly of carbon and oxygen and, in view of its high density, must be in crystal form. It is no longer a star in any recognizable sense of the word but has become, for all intents and purposes, a truly weird sort of *planet*!

Because of its density and elemental composition, it turns out that more than its position in orbit around a pulsar is weird … but more about that anon!

Planets of "Normal" Stars

The discovery of pulsar planets, unexpected though it certainly was, did not upset our view of planetary systems too greatly. It simply demonstrated that planets could sometimes exist under weird circumstances and if these planets were themselves a little weird, well, that should come as no great surprise. But planetary systems orbiting "normal" stars—stars not too different from our Sun and the host of others belonging to the so-called Main Sequence of hydrogen fusing suns in the prime of their lives—will surely be broadly similar to our own. After all, that is what the treasured assumption that our home Solar System is not "special" implies!

But we were in line for a shock! On October 6, 1995, University of Geneva astronomers M. Mayor and D. Quueloz announced the discovery of a Jupiter sized planet in orbit around a rather Sun-like main-sequence star known as 51 Pegasi. The discovery had been made at the Observatoire de Haute-Provence. The planet had not actually been "seen" but high-resolution spectroscopy detected slight but regular wobbles in the star itself indicative of the presence of some perturbing object in orbit at very small distance. Because of the gravitational pull of the unseen body, 51 Pegasi would periodically approach and then recede from Earth and accurate observations of the slight Doppler shift in its light gave away the unseen object's presence. Moreover, by observing the period of the star's movements, both the mass of the unseen body and its distance from the star could be determined. This is where the surprises started. The mass of the perturbing body was computed as being just half that of Jupiter. This is far and away too small for it to be to be a dwarf star and so it could only be a large planet, probably of the gas giant variety similar to Jupiter. This was an interesting discovery, but the real surprise was its distance from the star. The planet was orbiting the star at the incredibly close distance of about 4,400,000 miles (7,000,000 km), taking just 4 days to make one full revolution of its orbit!

Given the prevailing wisdom that gas giant planets form relatively far from their host stars, the presence of such an object circling around its star in an orbit that would make Mercury seem remote was a difficult pill to swallow. Indeed, *any* planet in such an orbit was a difficult pill to swallow, although one made predominantly of iron seemed *just* credible to some (although the required size was a difficulty) and just such a composition was proposed by a few astronomers. Indeed, a minority opinion that what had been found may have been pulsations of the star—not a planet at all—was put forward but soon withdrawn by the very person who proposed it. Difficult to believe though its existence surely was, the planet of 51 Pegasi was real!

Of course, the situation would not be as bad if this was a cosmic freak—a "one off" or, at least, a very rare occurrence that presumably required highly unusual circumstances for its existence. That comforting alternative was, however, quickly dispelled. These "hot Jupiters", as they soon came to be known, began turning

Fig. 6.1 Artist's impression of the "hot Jupiter" HD200458b (Credit: NASA)

up time and again; so frequently in fact that for a while it began to seem as if they were the norm for extrasolar planets! We now know that they are not the norm and, in fact, really do constitute a minority. The reason for their prevalence among early discoveries was simply due to a selection effect. As the early planet hunters themselves were very aware, the variations in a star's motion can only be clearly detected if observed over a time equal to that of the planet's orbital period. If that period is only a couple of days, a short observing time is all that is required, but if the planet has an orbit equivalent to that of Jupiter, discovery can only come after years of monitoring. The early results of a program of monitoring will therefore only reveal a population of large planets in small orbits. Those more reminiscent of the Solar System will take far longer to find (Fig. 6.1).

Nevertheless, even as the years of monitoring passed by and planets other than hot Jupiters were added to the ever growing list of discoveries, the scene was one of bewildering variety that refused to turn up anything very reminiscent of our home system. The first multi-planet system was discovered in 1999 associated with the star Upsilon Andromedae. This continues to be one of the closest analogues of our own Solar System found to date, albeit quite different in the finer details as we shall see.

The discovery of worlds beyond the Solar System has now become commonplace. Several techniques have added to the tally. The radial velocity method by which the earlier planets were found has been joined by;

(a) Observations of transits. Observers in suitable locations on June 5/6 2012 were privileged to witness something that will not occur again for over a 100 years; a transit of the planet Venus across the face of the Sun. Although we did not notice it, the Sun's light was actually reduced by a miniscule amount during that event. It would have required great accuracy to detect it, but it did happen. Similarly, accurate measurements of the brightness of other stars can sometimes detect the passage of a large planet in transit across their discs. This method works best for large planets orbiting small stars where the percentage of the star's disk hidden by the planet, and therefore the dip in brightness itself, is greatest. It is also most effective where the orbital period of the planet is short and, of course, for it to work at all we have to be observing the extrasolar planetary system within its orbital plane. It is also best if the star's output is steady, so that the diminutions caused by planetary transits are not lost amongst its intrinsic fluctuations. Nevertheless, this method has turned out to be the second most productive in discovering planets. It has also enabled the diameter of planets to be estimated and, where the planets' masses have been calculated from radial velocity measurements, computation of their mean density made possible. This has shown that some of the "hot" worlds have a density even less than that of Saturn—the "float on water planet" of our own Sun's family. This in turn gives proof that the planets in very small orbits really are hot gas giants rather than the alternative hypothesis of super Earths or great balls of iron. Transit observations have also enabled spectrographic analysis of the atmosphere of some of these worlds through the detection of absorption lines of the major atmospheric constituents revealed in the spectrum of the host star. Although monitoring tiny fluctuations in a star's brightness might seem like searching for needles in haystacks, advanced amateur astronomers in this CCD age have the capacity to join this exciting hunt with a very real prospect of making a discovery.

(b) As a variant of this method, in cases where more than one planet orbits the same star, perturbations of the transiting planet by a second one can result in slight discrepancies in the timing of its transits. In this way, planets that do not themselves transit the host star, from Earth's perspective, have been found by accurate timings of their frequently transiting sibling. In another variation of this method, accurate timing of the eclipses of a binary star can betray the gravitationally perturbing presence of a planet orbiting one or both of the stars. Several planets have been found through this method as well.

(c) In some respects the most interesting method involves a consequence of relativity theory about which, ironically, Einstein himself vocally expressed disinterest! Relativity interprets gravity as curved space in the presence of massive objects and this curvature acts like a magnifying glass focusing the light of a more distant source. Thus, counter intuitively, when a star passes directly in front of a more distant one, the light of the latter is concentrated around the eclipsing star. Rather than hiding the more distant object, the event actually enhances it; and does so in a very specific manner. If the eclipsing (or "lensing" as it is more commonly called) star has a planet, this will contribute its own distinctive signature to the light curve of the lensing event. This microlensing phenomenon has contributed to the total tally of planets but by its very nature is a one-off event not amenable to follow-up observations of the type that would yield information about the orbits of any planets found. Nevertheless, information as to the mass of the planet and its distance from the star are available and the method has the advantage of being capable of the detection of planets at great distances as well as being sensitive to worlds orbiting far from their parent star. Some extreme microlensing events are also capable of detecting

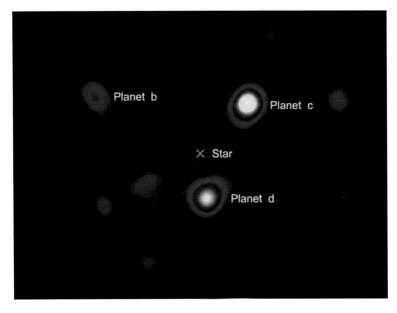

FIG. 6.2 Image of exoplanet 20100414 (Credit: NASA/JPL-Caltech/Palomar Observatory)

remarkably small planets; Earth-sized or even Mars-sized worlds are capable of detection in certain events.

(d) Accurate astrometric measurements of a star's proper motion—its slow movement over the years against the background of more distant stars—can detect any significant wobble in the start's motion due to the perturbation by a planet. Despite the early reports about which we have already spoken, no definitive discoveries have resulted from this method. It has, however, been employed to supply further data on planets found by other means.

(e) Features within curcumstellar disks, observed mainly in the infrared, suggest the presence of planets in some of these (Fig. 6.2).

These methods have netted a combined total of 777 confirmed planets, residing in 623 individual planetary systems as at mid 2012. Of these systems, 105 have more than one planet. These statistics will almost certainly have changed dramatically by the time you read these words, given that as many as 2,321 possible planets found by NASA's *Kepler* planet-finding space mission await confirmation at the time of writing. Some of these are undoubtedly spurious, but equally many will be confirmed and yet others added as the mission proceeds.

The first multiple planet system comparable to our own Solar System (i.e. other than the planets orbiting PSR B1257+12) was the Upsilon Andromedae system discovered in 1999. This system is now known to host four planets and, far beyond the planetary system at a distance of some 750 AU from the primary star, orbits a small secondary star of the red dwarf variety. The U Andromedae system is hardly a clone of our own Solar System, although by the standards of extrasolar planetary systems, it mimics it rather more closely than most. The main star, U Andromedae, is a yellow-white main sequence object a little larger and hotter than the Sun and also quite a deal younger; only about half the Sun's age in fact. It is located approximately 44 light years away and is visible with the naked eye at approximately fourth magnitude. The innermost of the system's planets—designated "b"—is a hot Jupiter orbiting the primary star at a distance of just 0.06 AU, completing one revolution of the star in just 4.6 days. This planet has a mass 1.4 times that of Jupiter. The second planet—"c"—has a mass of 13.98 Jupiters and orbits at a distance intermediate between that of Venus and Earth in our own planetary system (0.8 AU). It takes 241 days to complete a single orbit. The mass of this planet as given here is based on analyses published in 2010 and exceeds considerably earlier estimates which, however, only gave the estimated lower limit of its mass. If the newer results are upheld, this planet is actually verging on being a lightweight brown dwarf and may indeed be (or may once have been) fusing deuterium within its core! What we have may then be an object of duel personality; a planet at the upper limit of the spectrum of mass and, at the same time, a star of the smallest possible mass. However, even if it is—or was—fusing deuterium, it should still be listed as a planet (even if simultaneously as a brown dwarf star!) insofar as its position within a planetary system strongly suggests that it formed in the same way as its sibling worlds, even if its obesity caused it to (just) cross the line separating the largest gas giants from the smallest stars.

Project 1: Upsilon Andromedae

The principal star of the Upsilon Andromedae system is a relatively easy naked-eye object. Although there is, of course, no hint of its interesting companions, just pointing out the star with the first extrasolar multiple planet system to be discovered is sure to be a talking point at open nights for the local astronomy club.

The star is easily located about 10° east of the famous Great Andromeda Galaxy, M31. Upsilon and M31 are not too dissimilar in brightness although the galaxy, because it is an extended object, is less intense and more difficult to see in less than pristine skies. But once M31 is located, look east to the first rather bright naked-eye star. This is Gamma Andromedae, shining at a respectable second magnitude (magnitude 2.2 to be more precise). From Gamma, look approximately one third of the way back toward M31 along the imaginary line joining the two and you should find a relatively faint naked-eye star of fourth magnitude. This is Upsilon, the hub of the first multiple planetary system other than our own to be discovered in orbit around a more or less sunlike star.

This planet is sometimes said to lie within the star's habitable zone, however this should be taken with some caution as it is closer to the primary than Earth is to the Sun and the primary is also a little hotter than the Sun. It clearly receives more heat than Earth. Moreover, as a gas colossus verging on stellar mass, it is hardly a candidate for habitability no matter where it is placed in the system!

Planet d weighs in at 10.25 times the mass of Jupiter and orbits at 2.53 AU from the star, which in our system would place it in the middle of the asteroid belt. Planet e, the smallest of the quartet at 1.06 times the mass of Jupiter and, at 5.25 AU from the primary star, orbits at a similar distance to that of our own Jupiter. It requires 3,848 days to make a single orbit of the star. All of these planets have orbital eccentricities greater than Pluto. Moreover, they are not confined to the same plane like those in the Solar System.

The orbits of planets c and d, for instance, are inclined to each other at an angle of 30°. The orbit of c is also rather weird in so far as calculations reveal that it varies between circularity and maximum eccentricity in the relatively short time period of 6,700 years. Lack of detected dust beyond the realm of these planets also suggests that there is nothing equivalent to the Solar System's Kuiper belt. The presence of the red dwarf is suspected of having stripped away any Kuiper-type belt of small objects that may have started to form far from the primary.

Smaller planets (Earth sized?) may exist within this system, although the presence of four such massive worlds relatively close to the primary star makes their existence within the inner system very improbable. Small planets may exist out beyond planet e, however they would be cold and uninviting places with surface conditions very different from that of Earth.

Although there is no evidence one way or the other, the possibility that the large planets of this system may be accompanied by systems of moons is one raised from time to time and, because planet c is located within what some assessments define as the habitable zone, the question is inevitably raised as to whether this planet might be accompanied by a habitable (maybe even an *inhabited*) satellite. There are a number of factors to consider here. Given that the full habitability of Earth seems to rest on a knife edge maintained by a highly improbable set of factors, it seems rather unlikely that any moon of planet c would duplicate this precise situation. Notice though that we said "full habitability", meaning the set of conditions that allows the existence of complex life up to and including human beings. Habitability in a broader sense (for instance, conditions under which single celled organisms could exist) is much less restrictive as our earlier discussions have noted. A moon of planet c—or even of planets d and e—may well have as much chance of being habitable in this broader sense as, for instance, Europa. But whether a moon of planet c could sustain a complex surface ecosystem is a more difficult question. If it possessed an atmosphere as dense as Earth's, the extra heat received from the primary star might be enough to trigger a runaway greenhouse and the moon could turn out more like Venus than Earth. On the other hand, a smaller moon with a thinner atmosphere might lose much of this gaseous mantle to the stellar wind and photo dissociation, rapidly becoming unable to sustain bodies of liquid water on its surface. Moreover, a behemoth planet such as c would surely possess a powerful magnetosphere and because it orbits relatively near a fairly young and energetic star, this has just as surely trapped large quantities of energetic particles, bathing any nearby moons in powerful radiation. Not only is this detrimental from a biological point of view, but it would also combine with stellar radiation to denude the atmosphere of a moon. A sufficiently strong global magnetic field surrounding the moon might overcome this difficulty, as may a very wide orbit for the hypothetical moon (remember Callisto in the Jupiter system?). With reference to this last possibility however, it is likely that any moon of planet c would need to pursue a far wider orbit than that of Callisto to avoid the danger zone of this planet. And that raises yet another difficulty! If the moon orbits in the plane of the planet's orbit around the primary star, the moon's distance from that star necessarily changes and (if the moon's orbit is as wide as suspected), these changes would probably be sufficient to trigger large short-period fluctuations in climate. This must surely have some effect on habitability, in the narrower sense, and that effect will probably not be beneficial. This problem can nevertheless be avoided if the moon orbits its primary planet more or less perpendicular to the plane of that planet's orbit. This would permit the moon to move in a wide orbit, giving the planet and its radiation belts a wide enough berth, whilst maintaining a steady distance from the central star. The only variations in stellar distance would then be determined by the eccentricity of the planet's own orbit, which also goes through phases of significant variation in the case of planet c.

In short, the issue of life on a moon of Upsilon Andromedae c is a complex one and is, of course, entirely hypothetical at this stage. After all, we do not even know if any of the planets in this system have moons orbiting them!

Hot Jupiters and Other Toasty Worlds

The so-called "hot Jupiters" (or "roasters" as they are sometimes graphically described) were, as we have already seen, the first planets discovered in orbit around other sun-like stars. This, as we saw, was mostly due to a selection effect in so far as they are the fastest and easiest class of planets to find using the radial displacement method of planet hunting. They were also the ones which most astronomers least expected to find, as the only examples of gas giants previously known—those of our own Solar System—are relatively distant from the central star. It is not difficult in principle to understand how planets like Jupiter and Saturn can form at the distances of their present orbits as temperatures out there are low enough for the types of volatile material from which they accreted to be present. But how do Jupiter-and-Saturn-like planets form at the distances of the likes of 51 Pegasi b, Upsilon Andromedae b, or even Upsilon Andromedae c?

The short answer is: They can't! The so-called hot Jupiters like 52 Pegasi b and Upsilon Andromedae b and the "warm" Jupiters such as Upsilon Andromedae c must have formed at distances from their stars comparable to those at which Jupiter and Saturn formed from the Sun and then migrated inward, eventually coming to rest in tight orbits close to the central star. Interaction between planets may have caused some inward migrations, but the principal cause of this phenomenon is thought to have been the interaction between the accreting planets and the circumstellar disk from which they formed. The planet's gravitational attraction accelerates surrounding debris into larger orbits or even ejects it from the planetary system altogether, but as it does, part of the planet's momentum is lost and its own orbit slightly contracted. Repeated encounters and momentum transfers add up to a significant degree of inward migration; just how significant depending upon the amount of circumstellar debris and the length of time the disk persists. When the evolving star turns on strong stellar wind (which occurs during the T-Tauri phase of the life of a young star), the disk is dispersed, but before this happens the migrating planets within some systems may actually migrate so far inward as to plunge into the star itself. In other words, before a planetary system settles down into something resembling its final structure, at least one generation of planets may have ended up in the star. Nevertheless, if sufficient of the original circumstellar disk remains, others will form to take their place. The ones that are there at the end are those in existence when the remnants of the disk are swept away by stellar wind or, in the case of less massive disks, when the supply of planet-making material simply runs out. Wherever in the system planets of the final generation have migrated will be the orbits in which they remain. A system of multiple gas giants such as Upsilon Andromedae can therefore result as the planets' inward migration ceases at differing stages. Planet b migrated far enough to become a hot Jupiter, planet c a "warm" Jupiter and planet d what we might call a "temperate" Jupiter. On the other hand, planet e, in common with Jupiter itself, probably did not migrate very far from the place of its birth. This, I hasten to add, is a very oversimplified account of the early evolution of this system and is given as a very generalized scenario only, not as a detailed account of this particular case. In reality, mutual perturbations by the planets of this system must have played a vital role in shaping the outcome, especially considering the colossal mass of planet c.

Because of their extreme proximity to stars, hot Jupiters have several characteristic features. They tend to move in orbits that are very nearly true circles and to have one hemisphere always facing the primary star. This means that one side of the planet is being perpetually baked by the star, but because of the dense atmosphere of these Jupiter-like worlds, this does not imply that one side is roasting hot while the other is freezing cold. What it does imply is that, in attempting to equalize the temperature between hemispheres, upper atmospheric winds of incredible velocity cross the twilight zones of these worlds. Winds of 4,400 miles (7,000 km)/h have been directly measured in the atmosphere of the hot Jupiter HD 209458 b. Because this planet transits its primary star, astronomers are able to measure the Doppler shift in the absorption line of carbon monoxide in its atmosphere, and this has given them direct measurement of the wind speed in the planet's upper atmosphere. Infrared measurements of the relatively small difference in temperature between the day and night sides of another hot Jupiter—HD 189733 b—have also enabled astronomers to calculate the approximate velocity of the wind required to perform this transfer of heat. Once again, supersonic velocities are found. The *minimum* wind speed required is around 4,500 miles (7,200 km)/h (similar to that of HD 209458 b) with possible maximum velocities as high as 22,000 miles (35,200 km)/h! These speeds apply to global convective wind systems, something like Earth's trade winds or, on a more local scale, sea and land breezes writ extremely large.

Presumably there are also smaller scale storms and vortices on these worlds, something like the "spots" on Jupiter and Saturn in our home planetary system. What sort of winds might occur there are simply beyond imagining.

If supersonic gales are not enough, what are we to say about rains of molten iron? Temperatures in the atmospheres of hot Jupiters can be so high that iron is liquid and forms droplets of billowing clouds. As water droplets coalesce on Earth and fall as rain, so presumably do these cloud droplets, raining drops of molten iron into the lower reaches of the atmosphere of these frightening worlds.

Within the broader class of hot gas giants (notice that we have broadened the class beyond "Jupiters") two further subsets of planet can be distinguished; ultra-short period planets that whip around their stars so closely that their "years" amount to less than one terrestrial day and "puffy planets" or, as they are sometimes called "hot Saturns", not so much because of their mass as their low densities; densities which can be considerably less than that of Saturn itself. These planets are "puffy" because the combination of heating from the very close parent stars and internal heat generated by the planets themselves balloons them into large low-density balls of gas. Even though they may not be as lightweight as Saturn, they still tend to be rather less massive than some of the real heavyweights in the extrasolar planetary community. Planets larger than about twice the mass of Jupiter have enough gravity to hold in their girth against the expanding pressure of great heat. Incidentally, planets weighing in at several times the mass of Jupiter do not greatly exceed this planet's diameter; they are simply more compressed and denser.

One interesting puffy is HD 209458 b. This world orbits its sun at just one eighth the distance of Mercury from our Sun. It weighs in at a relatively light 0.69 times the mass of Jupiter, but is considerably more obese, occupying some 2.5 times that planet's volume. It is one of the planets that transits its star as observed from Earth's perspective and a surprising amount of detail has been uncovered because of this. For instance, it is known that its atmosphere contains methane, carbon dioxide and water vapor and that it sports some sodium in its outer atmosphere. It appears to have a cooler atmospheric layer as well as one of dark and very hot clouds, probably composed of vanadium and titanium oxides, although complex organic compounds of the type known as *tholins* are also a possibility. But the most surprising thing is the finding that it displays a comet-like tail some 124,000 miles (200,000 km) long. It is, in fact, slowly evaporating away into space. But it is not expected to disappear completely. Eventually, the atmosphere will all be swept away by the stellar wind, leaving something very interesting behind; the erstwhile core of the planet—a ball of rock and/or iron, probably around the size of Earth and known as a *chthonian planet*. Examples of these—the remnants of former evaporated hot gas giants—almost certainly exist and, indeed, two possible examples are already known as we shall see shortly.

Ultra-short period gas giants are more prone to ending up as chthonian planets. Typical giant planets orbiting at 0.02 AU or greater from their parent star may lose only 6–7 % of their mass, but a planet within 0.015 AU—an ultra-short period planet—can end up as a chthonian planet during the lifetime of its parent star. HD 209458 b, and similar puffy planets, are presumably more vulnerable and might become chthonian planets even at greater distances (although HD 209458 b still has a long way to go before reaching that stage) (Fig. 6.3).

Some ultra-short period planets may meet an even more dramatic end. An extreme representative of this class is WASP-18b; a world of ten Jupiter masses orbiting a star 1.25 times more massive than the Sun at the incredibly small distance of 1.9 million miles (3.1 million kilometers) or approximately 0.02 AU. Although observations of this planet have not extended over a sufficiently long period as yet, scientists suspect that it is spiraling inward even closer to the star and, if that is correct—and observations during the coming decade should tell whether it is—the planet is likely to be torn apart by the star's gravity within the next million years!

This appears to be happening already to another very close planetary companion of a sunlike star. This one is a very puffy hot gas giant known as WASP-12b. It has ballooned out to three times the radius of Jupiter and now has such a tenuous hold on its atmosphere that the parent star is pulling out a plume of the planet's gas and spiraling it down onto its photosphere. It is not uncommon for two close stars to interact in this way and this is, indeed, how

290 Weird Worlds

Fig. 6.3 Artist's impression of exoplanet Kepler 20e; a planet similar to Earth in size but too hot to be habitable (Credit: NASA)

cataclysmic variable stars such as nova generate their outbursts. But here we have the same process working in miniature with the "gas donor" being a bloated planet instead of a bloated star!

Earlier, mention was made of those strange planetary remnants known as chthonian planets. These are mostly hypothetical, although the Galaxy is old enough for some to have formed and two entries in the catalog of extrasolar planets are actually suspected of being possible candidates. These two worlds are COROT-7b and Kepler-10b, both relatively small worlds of incredible heat. The first of these is thought to be about 1.58 times the mass of Earth, although the exact mass has not been determined and there is still considerable leeway in the estimates. What is not in doubt however is its extreme proximity to its parent star; just 0.0172 AU or one twenty-third the distance of Mercury from the Sun. This makes the sunward hemisphere as hot as the filament of a tungsten bulb and it is suspected that the surface of this world is covered by an ocean of molten rock. This "lava planet" may even be surrounded by a very tenuous atmosphere … composed of vaporized rock!

The second planet is estimated to have a mass between 3.3 and 5.7 times that of Earth and to be about 1.4 Earth radii across.

It also orbits its star extremely closely at just 0.01684 AU with an orbital period—a "year"—of just 20.1 h. The surface of this world is hot enough to melt iron.

Both of these planets have been listed as "terrestrials" or "super terrestrials" but their extreme orbits suggest that they may really be the remnant cores of what were once hot Jupiters of ultra-short orbital periods, in other words, true examples of chthonian planets.

But one of the oddest of the very hot planets must surely be the "rock-comet world" discovered by the *Kepler* space telescope and announced in early 2012. At the time of writing, the interpretation given here for the odd-ball *Kepler* results has not been confirmed but (bizarre though it appears) it does seem to be the most reasonable solution.

Kepler observations of a star cataloged simply as KIC 12557548 revealed a fluctuation in its light that followed a very regular basis. Every 15.685 h *precisely*, the star dimmed. The star itself does not seem unusual; a quite well-behaved object just a little fainter than our own Sun. Moreover, the exact regularity of the dimming episodes strongly implies an orbiting planetary body transiting the star, similar to the increasing number of planetary transits being discovered by the *Kepler* probe and by other planet searches. However, there is one big difference between these episodes and the usual planetary transits. Although the KIC 12557548 episodes are highly regular in their timing, they are all over the place in the degree of dimming observed. Sometimes the star dims significantly, other times much more modestly. This is difficult to explain if they are being caused by a large planet passing in front of the star, for the simple reason that planets do not expand and shrink dramatically over a time span of hours or days! A pulsating planet like this seems just a little *too* weird to contemplate, even in this Universe of weird worlds!

One hypothesis put forward suggested that the transits might be caused by a giant planet in a variable orbit, albeit maintaining a highly regular orbital period. The suggestion is that sometimes the entire planet transits the star while only partially transiting the disk at others. Full transits result in significant dimming while grazing ones cause just a slight change in the star's brightness. The problem with this explanation however, is that a variable orbit of this type requires a second planet gravitationally perturbing the first. That in itself is not a problem per se and is, indeed, a rather common occurrence in the Universe, but if the perturbations in this instance are sufficient to move the transiting planet around quickly enough to account for the observations they would also be strong enough to seriously alter its orbital period and therefore change the timing of the dimming events. Thus far, the very precise periodicity of the dimming episodes has been maintained; in fact, the period has been maintained to within an accuracy of one in 100,000. A perturbing second planet does not appear to be the solution of this mystery.

Another explanation for these strange findings involves contamination of the star's light through the presence of a nearby binary star. Nevertheless, no independent evidence for such a star has been forthcoming.

The most popular—and most likely—hypothesis is that a small and rocky planet about the size of Mercury is orbiting the star at such close range that it is literally boiling away into space. From the orbital period (equal to the time between transits) it is inferred that the planet orbits just 0.01 AU from the center of the host star. An object that close to an approximately Sunlike star experiences a surface temperature about one third that of the solar surface; hot enough for the main constituents of rock to be vaporized. The surface of the planet is sublimating into gas which its weak gravity cannot retain at such small stellar distances. The planet is therefore thought to be surrounded by a cloud of gas, just as an icy comet nucleus becomes surrounded by a gaseous cloud as it passes through the inner Solar System. The planet, in effect, is a comet—a "rock comet"—and may even sport a long tail pointing away from its host star. Unlike a typical comet in our own System however, this object is stuck in a very close circuit of the star and cannot escape again into the cool reaches of space. Sooner or later, it will evaporate away completely.

The diming events are variable in depth because the gas cloud surrounding the evaporating planet is not constant. Sometimes there will be outbursts of gas, temporally making it more nearly opaque and causing a considerable dimming of the star's light. At other times, it will be less extensive and/or thinner and the dimming correspondingly less pronounced.

Lonley Planets and Fereezing Worlds

The technique of microlensing has brought to light some very weird worlds which don't really fit into any known categories of astronomical object. They are like planets, but if a "planet" is defined as a sub-stellar object orbiting a star, they don't fit the definition in one very important respect; they float freely through interstellar space!

Various names have been proposed for these objects—from the formal *sub-brown dwarfs* to a collection of more picturesque suggestions (some of them originating in science fiction stories penned long before any such objects were found) such as, *nomad planet*, *orphan planet*, *interstellar planet* and *rogue planet*. Personally, I like the sound of a rather romantic name given to them in a 1950s sci-fi television series; *gypsy moon*, although the term "moon" is hardly accurate and the title may no longer be considered politically correct! *Vagabond planet a*lso sounds good and I don't think has been suggested to date.

But whatever we call them, there are plenty out there in the Galaxy! A microlensing survey of 50 million stars carried out by the Microlensing Observations in Astrophysics (MOA) and Optical Gravitational Lensing Experiment (OGLE) and published in 2011 found 474 microlensing events of which ten were brief enough to have been caused by Jupiter size planets sans associated star. That may not seem very many, but the team concluded that to achieve this frequency, the number of Jupiter size free floating objects in the Galaxy needed to outnumber visible stars by almost two to one! Other researchers are not nearly so conservative, with estimates ranging as high as 100,000 to every star in the Milky Way.

These bodies (dare we call them *gypsy planets?*) are probably a mixture of true planets expelled from their solar systems through close encounters with their siblings and genuinely substellar objects that formed in the manner of stars and brown dwarfs, but simply lacked the mass to become either. It is hardly likely that the first variety consists only of Jupiter type planets. Gypsy Earths must surely be out there as well. Some of these may even be complete with their original moons as expulsion from a planetary system will not necessarily tear away a planet's satellites.

What would the scene be like on a gypsy Earth? The immediate thought is of a freezing world of eternal darkness, unless the world happens to be passing close to a star; an unlikely event to be sure. But while the dark part is certainly true, the cold may be a lot less than we might suppose. Studies have shown that even a body no larger than Earth, if far from any stars, could maintain a rather thick hydrogen and helium atmosphere sufficient to retain heat from the decay of radioactive elements and even tidally generated heat from a large retained satellite. Geothermal activity may even permit the existence of oceans and—who can say?—some form of life! We can hardly imagine very advanced life on these bodies, but given their estimated numbers, if even a minority support single-celled organisms the total Galactic census of such organisms would be enormous.

If we think that a wandering world drifting alone through interstellar space is the epitome of loneliness, then spare a thought for *intergalactic* planets. Now they would really be isolated worlds! We have no direct knowledge of such objects, but there are sound reasons for believing that they do exist. It has been known for quite some time that in certain circumstances a binary star system wandering too close to a supermassive black hole can be torn apart in such a way that one of the component stars is pulled into orbit around the black hole while the other is shot outward at such high velocity as to escape from the host galaxy altogether and go off into the void of intergalactic space. If the ejected star has planets, they will be carried along with it, but more recent studies at the Harvard-Smithsonian Center for Astrophysics and Dartmouth College have shown that even planets orbiting the star that does not get expelled from the galaxy may still meet with a similar fate. In certain circumstances, these could be stripped away from the star and ejected at such high velocity as to end up alone in the depths of intergalactic space. Planets ejected from double star systems—with or without their parent star—typically shoot through space speeds of 6–9 million miles (10–15 million kilometers) per hour but even higher speeds can be reached under favorable circumstances; up to around 30 million miles (50 million kilometers) per hour in fact. It is this class of extremely high-velocity worlds that end up escaping from the confines of the galaxy altogether.

Imagine the sky as seen from a planet deep in intergalactic space. No individual stars would be visible and, of course, no planets. Even is the planet had a moon, that too would be dark as pitch as nowhere would there be sufficient light to reflect back to its primary planet. From Earth, our host Galaxy extends in a great arc across the sky, its two Magellanic satellite galaxies are quite bright, the Great "Nebula" in Andromeda is a relatively conspicuous naked-eye object on a good night, the Triangulum galaxy is faintly discernible sans optical aid and one or two other more distant ones have been marginally detected with very keen naked eyes. The naked-eye galaxies of Andromeda and Triangulum, plus the Magellanic Clouds are all members of the small cluster of galaxies to which our Milky Way belongs. A starless intergalactic planet that had been expelled from the Milky Way, Andromeda or Triangulum would see its home galaxy initially brightly but slowly fading over time. The other major galaxies of the cluster would appear as dim fuzzy patches in an otherwise completely blank and incredibly black sky. Eventually, after eons of time, even these few spots of light would fade into the distance. No stars, no galaxies would remain. Just a void of the deepest blackness. Nothing in the Universe could be more alone!

Coming back just a little closer to cosmic company, planets in wide orbit around small and faint stars are also high on the list of cold, dark and lonely places. Such a world is the superterrestrial planet OGLE-2005-BLG-390L b discovered, by the microlensing technique, in orbit around a very dim red dwarf star. This planet is thought to be a ball of rock and ice about 5.5 times the mass of Earth orbiting its tiny sun at a distance of 2.6 AU and requiring some 10 years to make one full revolution. That distance would place it in the asteroid belt of the Solar System, but because the primary star is so dim, conditions on this world are a lot colder than on any of our asteroids. The average surface temperature is just 50°, or thereabouts, above absolute zero; a truly "plutonic" temperature. This is the coldest extrasolar world discovered at the time of writing. A cold day at the poles of Mars is balmy by comparison!

Diamonds in the Sky …

The chemistry of the Galaxy is evolving over time. Stars of each succeeding generation synthesize more heavy elements from lighter ones and, when reaching the end of their lives, return some of this heavy-element enriched material back to the interstellar medium. In the Galaxy, when ashes return to ashes and dust to dust, the ashes and the dust carry a trace of the stars' life's work back into the interstellar medium and each new generation of stars consequently forms from material having a different proportional elemental makeup than the last. Stars from generations younger than the Sun form from material that has been more enriched with carbon than that from which older stars formed (other things being equal) and this has the potential to lead to some pretty weird planets in orbit around them.

One such planet—one which we met earlier—is known by the unassuming designation WASP-12b. It is a giant—its estimated mass is about 1.4 times that of Jupiter—and it whirls around its parent star so closely that it completes each trip in a mere 26 h. But this is not the weirdest thing about the planet. There is evidence that this is a carbon-rich world. Beneath its scorching atmosphere a layer of graphite is thought to exist and, as heat and pressure increase with increasing depth, at some level this graphite is thought to give way to carbon in its crystalline state: To, in a word—*diamond*! Closer to the core, it is even possible that this could yield to even more exotic forms of the element, but for most of us, a world made largely of diamond is exotic enough.

Another recently discovered planet initially thought to be a rocky super-Earth is also now believed to be a carbon world with a diamond heart. This is the planet 55 Cancri e, estimated to have a diameter twice that of Earth but a mass eight times as great and to be composed mostly of carbon. As much as one third of its mass is thought to be in the form of diamond (2.7 Earth masses of diamond!) with a mixture of both diamond and graphite constituting its surface.

… Planets of Steel …

Recent research into the evolution of carbon rich worlds suggests that carbon planets about 15 times more massive than Earth could have *half* of their bulk in the form of diamond. Not only that, but as iron sinks to the cores of such worlds, the combination of iron and carbon could literally give them cores of steel. We might even speculate further and imagine what might result from the violent impacts that some of these worlds must surely sustain during the typically chaotic early period of their solar systems. Do naked steel cores or steel cores surrounded by diamond mantles comprise a small percentage of the planets within our Galaxy? The answer may indeed be "Yes" and, as carbon increases in the interstellar medium, they may become more common as the Galaxy ages.

Now, diamond planets may be interesting, but they are not likely to be a girl's best friend. In fact, they are not likely to be anyone's best friend! Diamonds rapidly lose heat, so we would expect diamond worlds to likewise lose their internal heat quickly and therefore lack the benefits that molten metallic cores bring. There would be no plate tectonic activity re-cycling the surface and no global magnetic fields to shield against energetic particles emitted by the parent star. They may sound fascinating, but diamond planets emerge as inert, cold and lifeless places. Not at all inviting really!

A 10,000,000,000,000,000,000,000,000,000,000 Carat Diamond!

Let us now return to a planet that we visited earlier in this chapter; PSR J1719-1438b, the planet that was once a star orbiting very close to the millisecond pulsar PSR J1719-1438. We said that this planet is believed to be predominantly composed of carbon and oxygen and is so dense that it must be crystalline in nature. Because of the high carbon content, much of the mass of this object probably exists in a state that could best be described as diamond. As one writer expressed it, the planet is essentially a 10^{31} carat diamond!

Because of the unusual method of formation, this is a different kind of diamond planet to the hypothesized steel-core-diamond-mantle variety and once again demonstrates just how exotic and far removed from our familiar Earth planets can be.

... And Still More Oddities!

Compared with the above, nothing found in orbit around another star or free floating in space should surprise us; but the oddities just keep on coming! It would be no surprise if even weirder worlds than anything mentioned here turned up before you read these words, so nothing said here even pretends to cover the whole spectrum of weirdness. Let us just round off this preliminary survey of extrasolar oddities with a few more examples of what has already been found and simply leave a question mark over what might be found in the future.

- **Brown dwarf planet**. Not only fully-fledged stars have planets in orbit around them. Brown dwarfs can as well, as was discovered in 2004 when infrared imaging found a body having just 3.3 times the mass of Jupiter orbiting the brown dwarf 2M 1207 at a distance from the primary of 41 AU. This was, incidentally, the first time that an extrasolar planet was discovered by direct imaging and the discovery was possible only because of the amount of heat (presumably arising from gravitational contraction) radiated by this peculiar planet. If a brown dwarf is thought of as being somewhere between a full star and a planet, this planetary body (now designated as 2M 1207b) could be said to lie somewhere between a planet and a moon! More recent monitoring of this system has also discovered a disk surrounding the brown dwarf, presumably material from which the giant planet accreted.
- **A Turkish bath planet**. We could be excused for calling our Earth a waterworld. After all, water covers some 71 % of its surface. Yet, in terms of its total bulk Earth is really pretty dry. Water accounts for a mere 0.02 % of its entire mass. By contrast, the planet designated GJ 1214 b appears to have most of its mass in the form of water; and a good deal of this is in some pretty weird forms!

    This planet is a superterrestrial having some 2.7 times the diameter of Earth and orbits just one and one quarter million miles (two million kilometers) from a dim red dwarf star about 40 light years from the Solar System. It completes one orbit of the star every 38 h. Despite the feeble nature of its primary star, its proximity means that the planet is a hot one, with temperatures several times higher than the boiling point of water at Earth's sea level. Because this planet transits its star, astronomers have been able to determine its diameter and, because it is so close to the star, they have also determined its mass from the wobble its gravitational attraction induces. Combining the two, the planet's density has been computed at a rather odd 1.9 g/cm^3. This is "odd" because it is too high for gas but not high enough for rock. What fits it nicely though is a mixture of water and rock. This model receives at least partial confirmation by *Hubble Space Telescope* observations of near-infrared spectra whilst a transit of the star was in progress. A team led by Zachory Berta of the Harvard-Smithsonian Center for Astrophysics used *Hubble* to monitor the planet's parent star while a transit was happening and again when one was not. By comparing the two spectra, the absorption of the star's light by the planet's upper atmosphere could be detected and the main constituent of the atmosphere identified. It seems that the chief single ingredient was water vapor, making up around 50 % of the atmosphere as against just 1 % of Earth's. With a temperature of several 100° and a 50 % water-vapor atmosphere, this planet makes a hot and humid day in one of Earth's tropical rain forests seem quite pleasant. Almost certainly, the planet's heat gets even more extreme with increasing depth, but it is thought that as the pressure also goes up the water vapor at greater depths turns into liquid, forming a bizarre and super hot alien ocean. At still greater depths within the planet—down toward its core—extremes of temperature and pressure are hypothesized to turn the water into phases totally unlike anything here on Earth—a literally unearthly superfluid and, at the most extreme depths, white-hot ice! Water ice as hot as white-hot steel, yet not melting!

In common with hot Jupiters, this planet probably formed further from the star and migrated inward.

- **Lilliput solar systems**. Imagine a tiny, faint, red dwarf star just 70 % more massive than Jupiter orbited by three small planets, one of which is about the same size as Mars. Imagine, further, that each of these orbits the tiny star at distances within 0.6 % of the Earth's distance from the Sun; so close that, despite the faintness of the star, the surface of the innermost world is at least as hot as Venus. Just such a system has been found! John Johnson of NASA's Exoplanet Science Institute describes it as being "more similar to Jupiter and its moons in scale" than other known systems of planets. Another, somewhat less extreme counterpart, designated Kepler-33 b also harbors a remarkably compact grouping of five planets—all orbiting closer to the parent star than Mercury is to the Sun! Maybe Lilliput planetary systems are not at all rare.

Giant steel ball-bearing worlds encased in diamond shells, diamonds half as large as Jupiter, worlds where water ice has the temperature of white-hot metal, systems of planets that could fit inside the satellite systems of Jupiter or Saturn, evaporating planets, worlds washed by lava oceans, "planets" wandering as celestial vagabonds through interstellar and even intergalactic space, worlds just a few tens of degrees above absolute zero and planets verging on the mass of lightweight stars. The diversity of planets already discovered or strongly supported by theoretical considerations is surely amazing; and we have the feeling that so far we have scarcely scratched the surface of the extrasolar planetary zoo. But event the small scratch on the surface that we have made is enough to show how parochial our vision has been until very recently. Only since the final decade of the Twentieth Century have we come to acquire some appreciation of the wonderful variety within this Universe in which we find a home. What amazing, strange and downright weird discoveries await us in future years? Time alone will tell, but we can be sure that they will be exciting!

Name Index

A
Adams, J., 152
Alcock, G., 124
Alexander, R., 63
Alexandre, C.M., 12
Allen, M., 263
Antoniadi, E., 6
Aristotle, 265
Asimov, I., 177, 244
Attree, N., 88

B
Barr, E., 185
Berta, Z., 299
Best, E., 82
Bland, P., 143
Bok, B., 20, 26
Bok, P., 20, 26
Borgonie, G., 61
Bouvard, A., 104
Brown, M., 155

C
Campbell, B., 270
Cassini, G., 66, 85
Cedillo-Flores, Y., 55
Challis, J., 104
Charles I of France, 38
Charlier, C., 125
Chodas, P., 164, 165, 167
Clark, R., 261
Clarke, J., 60, 61
Coleridge, S., 50
Colombo, G., 10
Comas Sola, J., 243, 244
Copernicus, N., 265–269
Correia, A., 12
Cunningham, C., 121

D
d'Arrest, H., 105
Dawes, W., 111
Diniega, S., 55
Dunham, E., 100

E
Elliot, J., 100
Encke, J., 85, 86

F
Fielding, G., 180
Firsoff, V., 188
Flamsteed, J., 91
Frail, D., 272

G
Galileo, G., 71, 81, 266
Galle, J., 105, 229
Gassendi, P., 5
Gastineau, M., 12
Gauss, K., 118, 123
Gounelle, M., 143, 144
Greenacre, J., 185

H
Haas, W., 185
Hall, A., 188
Hartnup, J., 111
Hauri, E., 182
Hencke, K., 124
Herbert, F., 255
Herschel, J., 242
Herschel, W., 91, 99, 121, 183
Hindley, K., 139
Hirayama, K., 128

Name Index

Hooke, R., 66
Horst, S., 258
Huth, J., 124
Huygens, C., 242

J
Jacob, W., 270
Johnson, J., 300
Jones, E., 60, 61

K
Kant, I., 269
Karrer, S., 49, 50
Keeler, J.F., 85
Kepler, J., 5, 19
Kiess, C., 49, 50
Kiess, H., 49, 50
Kingsborough, E., 82
Kirkwood, D., 128, 129
Kowal, C., 148
Kozyrev, N., 185
Kreutz, H., 164
Kuiper, G., 244

L
Laplace, P-S., 269
Laskar, J., 12
Lassell, W., 105, 110
Le Verrier, U., 152
Lemonnier, P., 91
Levin, G., 58
Lineweaver, C., 60
Lomonosov, M., 25
Longfellow, H., 38
Lorenz, R., 257
Lovejoy, T., 166
Lowell, P., 39, 118, 128, 152
Lyne, A., 270

M
Mandt, K., 253
Marsden, B., 196
Mayor, M., 277
McKay, C., 260, 262
McNaught, R., 160
Miller, J., 58, 59
Mink, D., 100
Moore, J., 249

Moore, P., 10, 26, 40, 98, 106, 118, 180, 185, 188, 189, 196
Morabito, L., 201
Moulton, F., 270
Murray, C., 129
Musgrave, I., 16

N
Nasmyth, J., 111

O
Ocampo, S., 180
Olbers, H., 123, 124, 126, 127
Onstott, T., 61
Oort, J., 157
Opik, E., 157
Origen, 37

P
Pascucci, I., 63
Piazzi, G., 122
Pickering, W., 152, 184
Prokopiy, 142, 143
Ptolemy, 265, 267
Pythagoras, 5

Q
Quueloz, D., 277

R
Reitsema, H., 111
ReVelle, D., 141

S
Sahecki, T., 40
Schiaparelli, G., 6, 39
Schroeter, J., 123
See, T., 270
Sekanina, Z., 164
Smith, H., 260
Spurny, P., 143
Straat, P., 58
Strobel, D., 261
Stubbs, T., 187
Swedenborg, E., 269
Swift, J., 188

T
Thomas, D., 271
Tobie, G., 254
Tokano, T., 257
Tombaugh, C., 152, 153

V
van de Kemp, P., 270
von Gruithuisen, F., 180

W
Walker, G., 270
Wetherill, G., 141
Wisdom, J., 129
Witt, G., 125

Wolf, M., 125
Wolszczan, A., 272

Y
Yang, S., 270
Yelle, R., 258

Z
Zuber, M., 15
Zupi, G., 4

Subject Index

A
Almagest, 267
Alphons, 185, 186
Aphrodite Terra, 21
Apollo, 5, 6, 126, 178, 181, 182, 184, 186, 187
Asteroids
    colors of, 138, 214
    families (*see* Hirayama families)
    named
        Achilles, 125, 126
        Amor, 126, 127
        Apollo, 126
        Astraea, 124
        Aten, 128
        Ceres, 122, 123, 130, 133–138, 147, 173
        Damocles, 159–161
        Eros, 125
        Fortuna, 138
        Gefion, 137
        2009 HC82, 161
        Hermes, 6
        Hidalgo, 127, 148
        Icarus, 127
        Pallas, 123, 124
        Patroclus, 126
        Patroclus I (Menoetius), 234
        Phaethon, 127, 138
        2006 RH120, 128
        2010 TK7, 126
        Vesta, 122–124, 130–133, 137
    Trojans, 126, 130

B
Barnard's star, 270

C
Caloris basin, 12, 13, 15
Cassini, 78, 79, 81, 84–90, 215–218, 220, 223, 242, 245–247, 250–253, 255, 261, 262, 264
Catastrophism, 114
Centaurs
    named
        Chiron, 149, 223, 224
        Echeclus, 149, 150
Chandrayaan-1, 182
Circumstellar disks, 286
Comets
    impact on rings of Jupiter, 70
    impact on rings of Saturn, 70
    named
        467, 165
        1106, 165
        1680, 164
        1843, 163–165
        1882, 163–165
        1945, 163, 165
        423 AD, 165
        214 BC, 165
        Elst-Pizarro, 147, 148
        Encke, 85, 139
        Hale-Bopp, 144, 158
        Halley, 144, 158
        Hartley, 144, 145, 158
        Hyakutake, 144, 158
        Ikeya-Seki, 163–166, 169, 170
        Ikeya-Zhang, 158
        LINEAR, 158
        Lovejoy, 166–170
        Pereyra, 170
        Shoemaker-Levy 9, 71, 90, 181
        Wild, 144
    sungrazing, 164, 165, 167–170

306 Subject Index

Copernican principle, 265–268
Cosmological principle, 268
Cryovolcanism, 216, 233, 247–249

D
DAWN, 122, 130–138
Deuterium, 43, 75, 144, 158, 282
Dwarf planets, 1, 3, 121–176, 209, 225, 236, 238, 258, 273, 298

E
Earth, 1, 65, 122, 175, 241, 265–269
Eris, 156, 173
Extra-solar planets
  designated
    55 Cancri e, 296
    COROT-7b, 290
    GJ 1214b, 298
    HD 189733b, 287
    HD 209458b, 287–290
    Kepler-10b, 290
    Kepler-33b, 300
    KIC 12557548b, 291
    2M 1207b, 298
    OGLE-2005-BLG-390Lb
    51 Pegasi b, 286
    PSR B1620-26b, 275
    PSR B1257+12b,c,d, 273
    PSR J1719-1438b, 297
    Upsilon Andromedae b,c,d, e, 285, 286
    WASP-12b, 290, 296
    WASP-18b, 290
  moons of, 298, 300
  types
    chthonian planets, 289–291
    diamond planets, 297
    eccentric Jupiters, 115
    Puffy planets (hot Saturns), 288, 289
    pulsar planets, 272, 276
    roasters (hot Jupiters), 115, 117, 277, 278, 285–293, 299
    rock-comet planets, 292
    steel planets, 297
    sub-brown dwarfs (vagabond planets), 293

F
Fireballs (meteors), 140
  in Taurid stream, 139, 141, 142
Forward scatter of sunlight, 70, 89

G
Gaia hypothesis, 35
Galileo, 71, 81–83, 103, 197–199, 206–208, 235, 266
Ganesa Macula, 248
Great Red Spot. *See* Jupiter
Great White Spot. *See* Saturn

H
Hadean era, 33
Hadley circulation, 29
Halicephalobus Mephisto, 61, 62
Haumea, 173
Hills cloud, 158
Hirayama familes of asteroids
  named
    Beagle, 128
    Ceres (Gefion), 137
    Eos, 128
    Kronos, 128
    Themis, 128
Hotei Orcus, 248
Hubble Space Telescope, 88, 95, 96, 102, 106, 112, 132, 134, 156, 237, 244, 299
Huygens lander, 246, 257

I
Ishtar Terra, 21

J
Jupiter
  atmosphere of, 287
  Great Red Spot, 66, 78, 79
  moons
    Adrastea, 71, 196, 197
    Amalthea, 71, 196, 197
    Ananke, 196, 197
    Callisto, 175, 196, 197, 202, 210–214, 245, 249, 285
    Carme, 196
    Elara, 196
    Europa, 196, 197, 199, 202, 205–210, 213
    Ganymede, 175, 196–199, 202, 210–214, 242, 245
    Himalia, 196
    Io, 72, 73, 177, 196, 199–205, 218
    Lysithea, 196
    Metis, 71, 196, 197
    Pasiphae, 196
    Sinope, 196

Subject Index   307

Thebe, 71, 196
ring system, 72, 114, 214

K
Kelvin-Helmholtz waves, 9
Kepler, 5, 19, 20, 282, 289–291, 300
Kirkwood gaps, 86, 128, 129, 146
Kraken Mare, 245
Kuiper belt, 118, 137, 151–171, 173, 223, 224, 230, 231, 235, 239, 284
Kuiper-belt objects (types of)
 cubewanos, 155
 plutinos
 twotinos, 155

L
Labeled Release (LR) experiment, 58
Lakes (of methane) on Titan, 244, 245, 251
Late heavy bombardment, 118
LEM. See Lunar Ejecta and Meteorites (LEM)
Life (possible on)
 asteroids, 138, 181
 Callisto, 213
 Ceres, 136–138
 Enceladus, 216–218
 Europa, 209–210
 extra-solar planets, 62, 279, 290, 298
 Ganymede, 211, 213
 Mars, 33, 34, 39, 40, 57
 Miranda, 226
 Moon, 209–210
 Pluto, 153
 Titan, 257–264
 Triton, 258
 Venus, 32–38
Long-runout landslides, 222
Lunar Ejecta and Meteorites (LEM), 186
Lunar Prospector, 179
Lunar Reconnaissance Orbiter, 182

M
Ma'adim Vallis, 46
Magellan, 20, 21, 25, 295
Makemake, 155, 156, 173
Maori legend of Saturn's rings, 82
Mariner 4, 40, 42
Mariner 9, 41, 45
Mariner 10, 10, 12, 13
Mars
 atmosphere of, 43, 44, 49, 50, 56, 57, 254

canals, 39, 49
dust fountains, 53
landslides, 54
methane, 43, 57, 59, 259
moons
 Deimos, 188–190, 192, 193
 Phobos, 181, 188–195
transient features on, 40
wave of darkening, 49, 52
Mars Reconnaissance Orbiter, 46, 55
Maxwell Montes, 21, 27
Mercury
 atmosphere of, 3, 6, 8–10, 16, 17, 41, 210, 274, 288, 290
 ice on, 9, 17, 65, 115
 other names for, 5, 10, 17, 172
 phases, 4, 6
 tail of, 15–17
Messenger, 8, 12, 13, 15
Meteorites
 carbonaceous chondrites, 134, 136, 138–148, 158, 194, 195
 HEDs, (Howardites, Eucrites & Diogenites), 131, 133
 named
  Kaiden, 195
  Orgueil, 143–147
  Revelstoke, 144
  Veliky Vstyug, 142, 143
 possible microfossils in, 33, 136
Meteor streams
 Beta Taurid, 139, 142, 143
 Geminid, 127, 138, 195
 Meteorites from, 23, 32, 33, 56, 129, 131, 133, 134, 136, 138–148, 158, 179–181, 183, 186, 192
 Taurid, 139–143
Methane-based life on Titan, 244, 260
Methanopyrus Kandleri, 60
Microlensing, 280, 293, 295
Moon
 atmosphere, 73, 91, 162, 183, 186, 194, 203, 204, 209, 212, 232, 243, 247–257, 261, 285
 craters on, 181, 197, 207, 212, 215
 as Earth's companion planet, 176, 177
 formation of, 50, 71, 114, 115, 197, 211, 222, 227, 239, 274, 297
 ice on, 9, 182, 217, 222, 233
 transient phenomena observed on, 184, 186
 twilight phenomena observed by astronauts, 184

N

Neptune
    atmosphere, 92, 97, 105, 106, 108–110, 155, 231, 234
    composition of, 229
    magnetic field of, 8, 31, 73, 81, 108, 204, 206, 218
    moons
        Despina, 229
        Galatea, 113, 229
        Larissa, 229
        Naiad, 229
        Nereid, 225, 230
        Neso, 229
        Proteus, 229
        Thalassa, 229
        Triton, 105, 111, 229–234, 258
    ring system, 97, 99, 111, 114, 225
New Horizons, 71, 153, 237, 238, 240
Nice model of Solar System evolution, 181

O

Olympus Mons, 41, 45, 46, 56
Oort cloud, 117, 151, 156–163
Orcus, 173, 248

P

Pioneer Venus, 28
Pluto
    atmosphere of, 3, 154, 155, 162, 231, 232, 238
    composition, 153, 172, 237, 238, 240
    as Kuiper-belt object, 152
    moons
        Charon, 176, 236–240
        Hydra, 237, 240
        Nix, 45, 237, 240
        S/2012 (134340)1, 237
        S/2011 P1, 237

Q

Quaoar, 173

R

Rivers (of methane) on Titan, 245, 251, 253

S

Saltation, 256
Saturn
    atmosphere of, 3, 73, 76–78, 81, 88
    composition of, 83, 89, 148, 223, 258
    Great White Spots, 79
    moons
        Atlas, 215
        Calypso, 215
        Daphnis, 215
        Dione, 214–217, 245
        Enceladus, 81, 84, 88, 158, 159, 214, 216–219, 227, 245
        Epimetheus, 215
        Helene, 215
        Hyperion, 214, 219, 242
        Iapetus, 84, 214, 220–224, 249
        Janus, 215
        Mimas, 214, 216
        Pan, 61–64, 66, 85, 86, 196, 211, 215
        Pandora, 86, 87, 215
        Phoebe, 84, 214, 220, 222–224
        Polydeuces, 215
        Prometheus, 86, 87, 215
        Rhea, 214, 215, 242
        Telesto, 215
        Tethys, 214–216
        Titan, 80, 162, 199, 214
    ring system, 66, 70, 83–85, 88, 89, 91, 97, 99, 214, 215
Scattered disc objects, 155
Sedna, 159, 173, 174
STEREO, 16
Sturzstroms. *See* Long-runout landslides

T

Tartola Facula, 248, 249
Tharsis, 40, 46
Tholins, 83, 289
Transits
    of extra-solar planets, 62, 268, 273
    of Venus, 193
    visible from Ceres, 137
Trans-Neptunian objects
    2000 CR105, 159
    2008 KV42, 159
    2007 OR10, 173
    2006 SQ372, 159
T-Tauri phase of stellar evolution, 286
Tycho, 182, 185

U

Uniformism, 115
Uranus
    atmosphere, 91–94, 96, 97, 100, 105, 106, 108, 109, 154, 243
    axial tilt, 3, 9, 96, 98, 108, 109, 154, 176, 227
    composition of, 109, 148, 229
    magnetic field of, 98, 99
    moons
        Ariel, 226
        Cordelia, 101, 225
        Cressida, 225
        Cupid, 225
        Desdemona, 225
        Juliet, 225
        Mab, 102, 225
        Margaret, 225
        Miranda, 227, 228
        Nereid, 225, 230
        Oberon, 226
        Puck, 225
        Titania, 225, 226, 243, 244, 248, 250, 254–259, 261, 263, 264
        Umbrial, 226
    ring system, 70, 91, 97, 99, 101, 102, 111, 225

V

Valles Marineris, 45, 46, 133
Venera, 24, 27, 34, 252
Venus
    atmosphere of, 26, 30, 32, 34, 35, 60, 250
    life on, 36
    other names for, 37, 38
    phases of, 148, 191–193, 285
    polar vortices, 29, 251
    super-rotation of atmosphere, 28, 29, 250
Venus Express, 24, 25, 29, 31
Viking, 57, 60
Voyager 2, 104, 108, 231

W

"Weird terrain" on Mercury, 13
WISE, 126